# TEACHER GUIDE

## 9th–12th Grade

### Includes Student Worksheets

### Science

Labs

Supply List

Answer Key

# Master's Class Biology

**MASTER BOOKS**
— CURRICULUM —

**Author:** Dr. Dennis Englin

**Master Books Creative Team:**

**Curriculum Development:**
Kristen Pratt

**Chief Editor:**
Laura Welch

**Art Director:**
Diana Bogardus

**Editor:** Craig Froman

**Cover and Interior Design:**
Diana Bogardus
Jennifer Bauer

**Copy Editor:** Judy Lewis

First printing: April 2019
Seventh printing: March 2023

Master Books®, P.O. Box 726, Green Forest, AR 72638

Master Books® is a division of the New Leaf Publishing Group, Inc.

ISBN: 978-1-68344-151-9
ISBN: 978-1-61458-705-7 (digital)

Unless otherwise noted, Scripture quotations are from the New King James Version of the Bible.

Printed in the United States of America

Please visit our website for other great titles:
www.masterbooks.com

For information regarding promotional opportunities,
please contact the publicity department at pr@nlpg.com

**Author Bio: Dr. Dennis Englin** enjoys teaching in the areas of animal biology, vertebrate biology, wildlife biology, organismic biology, Biology, and astronomy. Memberships include the Creation Research Society, Southern California Academy of Sciences, Yellowstone Association, and Au Sable Institute of Environmental Studies. Dr. Englin's most recent publications include a text currently used in *Principles of Biology*. His research interests are in the area of animal field studies. He is a retired Professor of Biology at The Master's University in Santa Clarita, California. (B.A., Westmont College; M.S., California State University, Northridge; Ed.D., University of Southern California)

# Table of Contents

*Where Faith Grows!* _____ *Scriptural truth woven throughout each course*

# Course Description

This course provides important training and practice in developing skills involved in the study of biology, including observing and recognizing interactions and interdependencies of organisms in their natural environment, the use of a light microscope, dissection skills, and insights and recent advances in modern biology.

- ✔ Each chapter in the textbook has accompanying worksheets and quizzes in the teacher's guide.
- ✔ Each chapter in the text has a laboratory exercise that teaches particular skills, illustrates insights for the concepts studied and provides experience in preparing laboratory reports.
- ✔ Every 3 or 4 weeks the student is to take an examination.

Life is from God. When God removes life, an organism dies. In Darwin's day, some organisms appeared to be very simple when contrasted to larger, more complex organisms. Along with the blessings of modern tools to study life, it has become obvious that no life is simple. This makes sense because nothing that God creates is simple.

- ✔ The consideration of issues of biological origins and how the approach differs depending upon assumptions of the nature of life and reality. Whether one comes from a theistic or secular point of view makes a huge difference in the interpretation of origins. The course concludes with human origins that have huge implications as to whether or not we were created in God's image with an eternal destiny or the sum product of natural laws acting upon atoms and molecules.

## Features

| | | |
|---|---|---|
| ⊚ | **Target Level** | Designed for grades 9–12 1 Credit with Labs |
| ⊙ | **Flexible 180 Day Schedule** | 45–60 minutes per day 5 days per week |
| ⟫⟫ | **Open & Go** | Daily Schedule, Lab Supply List, Answer Keys |
| ✎ | **Engaging Application** | Worksheets, Labs |
| ⊞ | **Assessments** | Quizzes, Exams |

## Objectives

- ✔ Investigate the core concepts of historical and modern biology
- ✔ Become familiar with the meaning of key terms in biology
- ✔ Explore the fundamental concepts of cell biology and important recent developments
- ✔ Study the development of classical genetics and modern concepts in the expression of DNA and how it varies with age and changing environmental conditions
- ✔ Learn the coordination and maintenance of the parts of an organism

**OPTIONAL RESOURCE:** While *Biology Through A Microscope* is a book that can be used with any biology course, it also serves as a companion book in the *Biology: The Study of Life from a Christian Worldview* high school course. It contains full-color microscopic images of varied animals, insects, plants and fungi, and microorganisms, as well as detailed information for using the modern microscope in the classroom and discusses examples of stained and unstained slide samples, brightfield, darkfield, and phase-contrast microscopy. **Those who purchase this book would not have to use a microscope in order to fulfill the lab requirements.** A cross-reference sheet for related labs is included in the back of this teacher guide.

# Teacher Instructions for this Course

This teacher guide provides additional help in assisting students through the text and the laboratory exercises.

- **Reading:** The required reading in this course is very rigorous and detailed. For students who are having difficulty in reading the chapter or understanding some of the concepts, the second day can be used to either complete the reading and/or worksheet or to focus on specific parts of the chapter that may need review. It is vital that the student understand the concepts within the reading material to complete the course. The glossary, which contains the vocabulary words, begins on page 320 in the student book.

- **Vocabulary Words:** On the second page of every chapter in the student book, vocabulary words are introduced that are highlighted in that chapter's text and have brief definitions found in the glossary at the back of the book. Students are encouraged to either write these out on 3 x 5 cards or to create another useful means of reviewing these throughout their course of study. Comprehension of sometimes difficult terms and concepts is very important to completing a course in Biology or any other complex science study.

- **Labs:** Each chapter contains a lab, some of which also utilize a microscope, which is optional if you already have something that includes the microscope images needed for the course such as *Biology Through A Microscope*. See pages 7 and 8 for additional information on the labs and the supply list for items not readily available around the house or in a local store.

- **Worksheets:** The worksheets are important in indicating what the student needs to remember from the chapters and if they are ready for the quiz. The student needs to answer the questions in the worksheet. This is not a quiz, so the student can look at the answers after attempting them. The student needs to go back over the chapter to see why any questions were not answered correctly.

- **Quizzes:** After understanding the correct answers to the questions on the worksheet, the student is ready for the chapter's quiz. It should be taken without using the book or the chapter's worksheet. The course builds on concepts as it progresses. It is important for the student to master the concepts in each chapter.

- **Exams:** Every 3 or 4 weeks, the student will take an exam. In preparation for the exam, the student can choose as needed to reread the chapters covered on the exam or review the questions and answers on the worksheets and chapter quizzes. Copies of the worksheets and quizzes can be made if the student wants to do them again to prepare for the exams. The student is ready for the exam after studying the chapters, worksheets, and quizzes. The teacher should grade the exam and record the grade.

Remember, it is normal to do better in some areas than in others. Learning is more than memorizing. Some of the questions depend upon recall, but some also depend upon reasoning skills. Through practice, these skills are developed. If the student gets discouraged, please be a source of encouragement.

The spiritual insights and Scripture references are important and not just a nice tack-on for this course. These references are deliberately given in context with the subject matter.

# Teacher Instructions for the Laboratory

## Biology Credits for Transcripts

This is a one-year course with two full semesters, helping the student fulfill one credit of biology, which includes the lab. High school transcripts will list the course as Biology with Labs. If questions arise from state agencies or schools, they can be referred to the course content.

**NOTE:** *This information is given so that the teacher can come alongside the student in helping line up the necessary materials, overseeing the procedures where necessary, and evaluating the lab reports. Do not hesitate to ask someone with more background in biology or labs for advice in areas where you feel less confident.*

The labs for this course are designed to be done at home. I have not included anything in the labs that I would not want in my home. The specimens for dissection are not preserved in harmful compounds — even though they should still be kept out of the mouth, so be careful that young children are not able to get into the lab supplies.

It is also important to realize that the labs are included to help the student better understand the concepts presented in the course. This is not about perfection; things can and will go awry. It's the nature of scientific study and doing labs. Sometimes the student can do everything right, and for some unknown variable (materials, temperature, etc.), the experiment doesn't end up as intended. Encourage the student to do their best and PREP:

**P** – Prepare by organizing your lab supplies, so they are easily found to be used. Store them properly.

**R** – Review by reading the lab in your student book, and the lab pages in this teacher guide, before you begin. Lab pages in the student book also include an important paragraph at the end, giving context to the focus of the lab.

**E** – Expect to find the answer/process you are trying to find in the lab. Confidence in your process and yourself can help when things are hard to understand.

**P** – Proceed with a clear understanding of what you want to accomplish and how you need to fill out the lab pages. Be sure that complete sentences are used in the reports except where data are being recorded. This provides added writing experience and is clearer to someone reading the report.

- Students will use the "Laboratory" pages for taking notes, making general observations, and recording data.

- The "Laboratory Report" pages (see following examples) are for the student to write out their full observations, data, and conclusions. Any questions asked in the lab instructions are to be answered in the report.

- If the student completes the lab and does not get the expected result, he or she should write what happened, and also what they think should have happened. The student could also include his or her ideas on why he or she thinks the experiment turned out differently than expected.

**WARNING: As with any science course that includes laboratory exercises, some things can be potentially hazardous if not handled properly. Make sure to follow all instructions carefully:**

- Wear proper safety equipment when needed, including safety goggles/glasses.
- Keep small children away from where the labs are being conducted.
- Wash hands, surfaces, and equipment properly after each experiment.
- Make sure clothing and other household items and surfaces are protected from staining.
- Handle the microscope with care — always carry it with two hands.
- Do not place anything used for labs in your mouth. Some chemicals can be poisonous.

**Laboratory Report** (20 points possible)

*Living Things Observed in the Area*

At the Bozeman ponds (Bozeman, Montana) I see a body of water surrounded by tall willow trees and an occasional path leading down to the water. There are grasses farther on shore from the trees. The water is rippling with mallard ducks swimming across the water and on the shore. There is a lot of leaf litter on the shore and in the water from the trees.

*A Keystone Species*

The willow trees appear to be a keystone species. They stabilize the shoreline and provide shelter for the ducks. The leaves appear to be discolored perhaps with fungus and bacteria growing on them.

I cannot see any water insects or algae in the water, but the ducks are feeding on something in the water and these are typically fed upon by ducks. The ducks will deposit their waste in the water and on the shore which will provide nutrients for the algae, water insects, and shore plants. The ducks also appear to be a keystone species.

*Life Forms That Feed Upon Other Life Forms*

The ducks appear to feed on water insects and algae in the water.

*Life Forms That Are Eaten by Other Life Forms*

Water insects and algae appear to be eaten by the ducks.

*What decomposers are in the area?*

Bacteria and algae appear to be decomposing the leaf litter on the shore and in the water. This could be shown by the discoloration of the leaves which are slowly coming apart.

**Sample Biology Lab Reports** Some of the numbers have been changed from what is asked for in the lab so that the students have to do their own work.

## Laboratory 8

- [ ] Microscope (from the supply kit)
- [ ] Prepared microscope *Paramecium* slide (from the supply kit)
- [ ] Prepared microscope *Amoeba* slide (from the supply kit)
- [ ] Prepared microscope human blood smear slide (from the supply kit)
- [ ] Prepared microscope frog blood smear slide (from the supply kit)

## Laboratory 9

- [ ] Microscope (from the supply kit)
- [ ] Microscope Prepared Slide Onion Root Tip Mitosis (from the supply kit)
- [ ] Prepared Slide of *Ascaris* (roundworm) Mitosis (from the supply kit)

## Laboratory 10

- [ ] A notepad (something to write on) and a pencil
- [ ] 1 square foot of earth — **Optional**
- [ ] 1 square foot of screening or chicken wire — **Optional**
- [ ] Small bone (chicken wing or drumstick) — **Optional**

## Laboratory 12

- [ ] 5 2-inch-long segments of water plant (from pet shop or aquarium section of other stores, strands of algae found along a pond or stream will also work)
- [ ] 5 large test tubes (22 ml) (from the supply kit)
- [ ] 1 beaker (100 ml) (from the supply kit)
- [ ] Bright light source
- [ ] Distilled water
- [ ] Green, yellow, and red food coloring
- [ ] Phenol red pH indicator (from the supply kit)
- [ ] 1 drinking straw
- [ ] Timer — perhaps your smart phone
- [ ] Wax pencil (from the supply kit)

## Laboratory 13

- [ ] Distilled water
- [ ] 3 large test tubes (from the supply kit)
- [ ] 3 small test tubes (from the supply kit)
- [ ] 3 100 ml beakers (from the supply kit) of warm water
- [ ] Yeast (1 package of baker's yeast)
- [ ] ¼ teaspoon measuring spoon
- [ ] ½ cup measuring cup

- [ ] Glucose (also called dextrose) (from the supply kit)
- [ ] Sucrose (table sugar)
- [ ] Ruler (to measure mm, millimeters) (from the supply kit)
- [ ] Wax pencil (from the supply kit)

## Laboratory 14

- [ ] Microscope (from the supply kit)
- [ ] Microscope Prepared Slide Onion Root Tip Mitosis (from the supply kit)
- [ ] Microscope Prepared Slide Roundworm *Ascaris* (from the supply kit)
- [ ] 3 lengths of thick string 30 inches long
- [ ] 81 paper clips
- [ ] 14 strips of red colored paper ½ inch x 1 inch
- [ ] 16 strips of blue colored paper ½ inch x 1½ inch
- [ ] 14 strips of green colored paper ½ inch x 1 inch
- [ ] 16 strips of yellow colored paper ½ inch x 1½ inch
- [ ] 16 strips of white paper ½ inch x 1 inch
- [ ] Other colors of paper can be used, as long as you have 5 different colors

## Laboratory 15

- [ ] The 3 strands of string with the bases attached from laboratory 14
- [ ] 6 paper clips
- [ ] The 6 paper strips you set aside in laboratory 14 (AUGUGA) to attach to the paper clips for the third strand (mRNA)

## Laboratory 17

- [ ] Microscope (from the supply kit)
- [ ] Prepared slide of Muscle Types (from the supply kit)
- [ ] Prepared slide of Motor Neurons (from the supply kit)
- [ ] Antibiotic discs (ampicillin, erythromycin, neomycin, and penicillin) **Optional — for next week**
- [ ] *E. coli* live cultures — **Optional**
- [ ] Prepared Tryptic Soy agar media plates — **Optional**
- [ ] Sterile swab applicators — **Optional**
- [ ] Bleach — **Optional**
- [ ] Wax pencil (from the supply kit) — **Optional**
- [ ] 100 ml beaker about half full of bleach (from the supply kit) — **Optional**

## Laboratory 18

- ☐ Microscope (from the supply kit)
- ☐ Prepared slide of Fern Life Cycle (from the supply kit)
- ☐ Prepared slide of Moss Life Cycle (from the supply kit)
- ☐ Petri dishes with *E. coli* cultures from Laboratory 17 — **Optional**
- ☐ Petri dishes with other samples that you may have chosen to do — **Optional**
- ☐ Antibiotic discs (ampicillin, erythromycin, neomycin, and penicillin) (from the supply kit) — **Optional**
- ☐ Tweezers — **Optional**
- ☐ 100 ml beaker — **Optional**

## Laboratory 19

- ☐ PTC taste test paper (from supply kit)

## Laboratory 20

- ☐ Image of corn cob with kernels (seeds) produced from a cross of R/r Su/su x r/r su/su

## Laboratory 21

- ☐ Access to insects or pictures of insects

## Laboratory 22

- ☐ Microscope (from the supply kit)
- ☐ Prepared slide of *Ranunculus* root cross section (from the supply kit)
- ☐ Prepared slide of *Ranunculus* stem cross section (from the supply kit)
- ☐ Prepared slide of *Ficus* leaf cross section (from the supply kit)

## Laboratory 23

- ☐ Microscope (from the supply kit)
- ☐ Prepared slide of *Amoeba* (from the supply kit)
- ☐ Prepared slide of diatoms (from the supply kit)
- ☐ Prepared slide of *Euglena* (from the supply kit)
- ☐ Prepared slide of *Volvox* (from the supply kit)
- ☐ Sterile toothpicks (from grocery store)
- ☐ Yeast soaked in red food coloring (from grocery store)
- ☐ Mold (grown at home as prepared last week)
- ☐ Clean blank microscope slides (from the supply kit)

## Laboratory 24

- ☐ Microscope (from the supply kit)

- ☐ Prepared slide of earthworm cross section (from the supply kit)
- ☐ Prepared slide of *Planaria* (from the supply kit)
- ☐ Nitrile gloves (from the supply kit)
- ☐ Scissors with fine sharp tip (from the supply kit)
- ☐ Styrofoam dissection tray (from the supply kit)
- ☐ Safety scalpel #11 (from the supply kit)
- ☐ T-pins (from the supply kit)
- ☐ Earthworm dissection guide (from the supply kit)
- ☐ Earthworm specimen (from the supply kit)
- ☐ Grasshopper dissection guide (from the supply kit) — **Optional**
- ☐ Grasshopper specimen (from the supply kit) — **Optional**

## Laboratory 25

- ☐ Microscope (from the supply kit)
- ☐ Prepared slide of frog ovary (from the supply kit)
- ☐ Prepared slide of frog sperm (from the supply kit)
- ☐ Prepared slide of human skin (from the supply kit)
- ☐ Nitrile gloves (from supply kit or locally)
- ☐ Scissors with fine sharp tip (from the supply kit)
- ☐ Styrofoam dissection tray (from the supply kit)
- ☐ Safety scalpel #11 (from the supply kit)
- ☐ T-pins (from the supply kit)
- ☐ Frog dissection guide (from the supply kit)
- ☐ Frog specimen (from the supply kit)

## Laboratory 26

- ☐ A partially eaten chicken leg or similar piece of meat — **Optional**
- ☐ A piece of screen or chicken wire about 1½ feet x 1½ feet — **Optional**
- ☐ A plot of ground 1 foot x 1 foot and 2 inches deep — **Optional**
- ☐ Wet soil — **Optional**

**Note:** You may want to reference Lab 10, which utilized these same materials.

## Laboratory 27

- ☐ These depend upon what you have available. Be creative and adventurous.

## Laboratory 28

- ☐ Pictures of Neanderthal, Cro-Magnon, *Australopithecus afarensis* (Lucy), chimpanzee, and human skulls and skeletons as provided below.

# Teacher Instructions for Quizzes and Examinations

**Quizzes:** These are to be given at the end of the chapter study per the schedule of the lessons. The student should review the text of the lesson and the worksheets.

✔ Quiz answer keys are included in the back of this teacher guide.

✔ Have the student look up any questions that were missed and explain to you what the correct answer should be and why. This helps the student master the concepts that impact future chapters.

✔ The quizzes are multiple choice and matching (with few exceptions) to make grading easier.

✔ There are 28 quizzes with 15 points possible for each quiz. This gives a possible total of 420 points.

In science studies, an A and B are very good. C is average. D or F indicates the need for more focus, more practice, or more study. Future success is always possible with focus, study, and practice.

The customary grading scale is:

90%–100% is an A    80%–89% is a B    70%–79% is a C    60%–69% is a D    59% and lower is an F

This applies to each individual quiz. At the end of the course, the average of the quizzes is to be added to the average of the exams to give a final score graded according to this scale.

If a student misses more than 50% on a quiz, the quiz should be retaken after careful study. As always, how you decide to award points and the grading scale used is your choice.

One suggestion is to give the student back ½ point for each answer gotten correct the second time that was missed the first time. This could be done for up to 5 quizzes. This can be very helpful for students that get off to a slow start.

Also, a student's readiness for a study of this nature may depend more upon maturity rather than age. Always encourage your student but still hold the standard and try not to cut corners. That way, the student will have the assurance of being able to go on to further scientific studies and succeed.

There is no midterm or final examination because, by its very nature, Biology is comprehensive. The concepts learned earlier are used in the later lessons and labs throughout the course.

## Administering the Examinations

In the week of an examination, the student is to study the previous quizzes and the worksheets for the lessons covered on the exam. The exam is like an expanded version of a quiz. Each exam consists of 30 multiple choice or matching questions (with few exceptions). An examination is a sampling of the material and does not include every point covered in the lessons.

## Biology Credits for Transcripts

This is a one-year course with two full semesters, helping the student fulfill half a credit per semester of biology, which includes the lab. High school transcripts will list the course as Biology with Labs. If questions arise from state agencies or schools, they can be referred to the course content.

A high school transcript usually has 1 grade for science courses (lab and lecture combined), and so this would appear as 1 credit with labs in Biology. (Note that some states may calculate credits in a different manner.) This can be determined by making the quizzes and exams 75 percent of the grade and the lab 25 percent of the grade if you so choose. To find the lab grade take the total points earned from all of the labs divided by the total possible times 100. An example of finding the total grade is if the average of the quizzes and exams are 85 percent and the labs are 97 percent:

Quiz/Exam Average   85   x 3 =   255   + Lab   97   =   352   / 4 x 100 =   88% (B+)   Final Grade

# Grading Sheet

| Lesson | Quiz | Exam | Lab |
|---|---|---|---|
| Lesson 1 | _____ / 15 | | _____ / 20 |
| Lesson 2 | _____ / 15 | | _____ / 20 |
| Lesson 3 | _____ / 15 | | _____ / 20 |
| Lesson 4 | _____ / 15 | Examination 1 _____ / 30 | _____ / 20 |
| Lesson 5 | _____ / 15 | | _____ / 20 |
| Lesson 6 | _____ / 15 | | _____ / 20 |
| Lesson 7 | _____ / 15 | | _____ / 20 |
| Lesson 8 | _____ / 15 | | _____ / 20 |
| Lesson 9 | _____ / 15 | Examination 2 _____ / 30 | _____ / 20 |
| Lesson 10 | _____ / 15 | | _____ / 20 |
| Lesson 11 | _____ / 15 | | _____ / 20 |
| Lesson 12 | _____ / 15 | Examination 3 _____ / 30 | _____ / 20 |
| Lesson 13 | _____ / 15 | | _____ / 20 |
| Lesson 14 | _____ / 15 | | _____ / 20 |
| Lesson 15 | _____ / 15 | | _____ / 20 |
| Lesson 16 | _____ / 15 | Examination 4 _____ / 30 | _____ / 20 |
| Lesson 17 | _____ / 15 | | _____ / 20 |
| Lesson 18 | _____ / 15 | | _____ / 20 |
| Lesson 19 | _____ / 15 | Examination 5 _____ / 30 | _____ / 20 |
| Lesson 20 | _____ / 15 | | _____ / 20 |
| Lesson 21 | _____ / 15 | | _____ / 20 |
| Lesson 22 | _____ / 15 | Examination 6 _____ / 30 | _____ / 20 |
| Lesson 23 | _____ / 15 | | _____ / 20 |
| Lesson 24 | _____ / 15 | | _____ / 20 |
| Lesson 25 | _____ / 15 | Examination 7 _____ / 30 | _____ / 20 |
| Lesson 26 | _____ / 15 | | _____ / 20 |
| Lesson 27 | _____ / 15 | | _____ / 20 |
| Lesson 28 | _____ / 15 | Examination 8 _____ / 30 | _____ / 20 |
| Total Score / Percent | _____ / 420 = _____ % | _____ / 240 = _____ % | _____ / 560 = _____ % |
| | Quizzes _____ % + Examinations _____ % / 2 = _____ % | | _____ % |

Quiz/Exam Average _____ x 3 = _____ + Lab _____ = _____ / 4 x 100 = _____ **Final Grade**

| Date | Day | Assignment | Due Date | ✓ | Grade |
|---|---|---|---|---|---|
| | | First Semester-Second Quarter | | | |
| Week 1 | Day 46 | Review Lesson 5 and Lesson 5 Quiz | | | |
| | Day 47 | Review Lesson 6 and Lesson 6 Quiz | | | |
| | Day 48 | Review Lesson 7 and Lesson 7 Quiz | | | |
| | Day 49 | Review Lesson 8 and Lesson 8 Quiz | | | |
| | Day 50 | Take **Exam 2 (Lessons 5–8)** Pages 297–298 • (TG) | | | |
| Week 2 | Day 51 | Begin Chapter 9, Cell Division • Read Pages 86–88 to the last full paragraph • (BIO) Complete **Lesson 9 Worksheet 1** Page 77 • (TG) | | | |
| | Day 52 | Finish reading Chapter 9 • Pages 88 from the last paragraph–90 • (BIO) • Complete **Lesson 9 Worksheet 2** Page 78 • (TG) | | | |
| | Day 53 | Read Laboratory 9, Cell Division Pages 92–93 • (BIO) Start Laboratory 9 Pages 79–80 • (TG) | | | |
| | Day 54 | Conclude Laboratory 9 and Prepare **Lesson 9 Lab Report** Page 81 • (TG) • Review Chapter 9 for quiz | | | |
| | Day 55 | Complete **Quiz 9** Page 253 • (TG) | | | |
| Week 3 | Day 56 | Begin Chapter 10, Ecosystems • Read Pages 94–97 • (BIO) Complete **Lesson 10 Worksheet 1** Page 83 • (TG) | | | |
| | Day 57 | Finish reading Chapter 10 • Pages 98–101 • (BIO) Complete **Lesson 10 Worksheet 2** Page 84 • (TG) | | | |
| | Day 58 | Read Laboratory 10, Ecosystems Pages 102–103 • (BIO) Start Laboratory 10 Pages 85–86 • (TG) | | | |
| | Day 59 | Conclude Laboratory 10 and Prepare **Lesson 10 Lab Report** Pages 87–88 • (TG) • Review Chapter 10 for quiz | | | |
| | Day 60 | Complete **Quiz 10** Page 255 • (TG) | | | |
| Week 4 | Day 61 | Begin Chapter 11, Biomes • Read Pages 104–110 • (BIO) Complete **Lesson 11 Worksheet 1** Page 89 • (TG) | | | |
| | Day 62 | Finish reading Chapter 11 • Pages 111–117 • (BIO) Complete **Lesson 11 Worksheet 2** Page 90 • (TG) | | | |
| | Day 63 | Read Laboratory 11, Biomes Pages 118–119 • (BIO) Start Laboratory 4 Page 91 • (TG) | | | |
| | Day 64 | Conclude Laboratory 11 and Prepare **Lesson 11 Lab Report** Page 93 • (TG) • Review Chapter 11 for quiz | | | |
| | Day 65 | Complete **Quiz 11** Page 257 • (TG) | | | |
| Week 5 | Day 66 | Begin Chapter 12, Energy Capture — Photosynthesis • Read Pages 120–125 • (BIO) Complete **Lesson 12 Worksheet 1** Page 95 • (TG) | | | |
| | Day 67 | Finish reading Chapter 12 • Pages 126–127 • (BIO) Complete **Lesson 12 Worksheet 2** Page 96 • (TG) | | | |
| | Day 68 | Read Laboratory 12, Photosynthesis Pages 128–129 • (BIO) Start Laboratory 12 Pages 97–98 • (TG) | | | |
| | Day 69 | Conclude Laboratory 12 and Prepare **Lesson 12 Lab Report** Page 99 • (TG) • Review Chapter 12 for quiz | | | |
| | Day 70 | Complete **Quiz 12** Page 259 • (TG) | | | |

| Date | Day | Assignment | Due Date | ✓ | Grade |
|------|-----|------------|----------|---|-------|
| Week 6 | Day 71 | Review Lesson 9 and Lesson 9 Quiz | | | |
| | Day 72 | Review Lesson 10 and Lesson 10 Quiz | | | |
| | Day 73 | Review Lesson 11 and Lesson 11 Quiz | | | |
| | Day 74 | Review Lesson 12 and Lesson 12 Quiz | | | |
| | Day 75 | Take **Exam 3 (Lessons 9–12)** Pages 299–300 • (TG) | | | |
| Week 7 | Day 76 | Begin Chapter 13, Energy Release — Respiration • Read Pages 130–134 • (BIO) <br> Complete **Lesson 13 Worksheet 1** Pages 101 • (TG) | | | |
| | Day 77 | Finish reading Chapter 13 • Pages 135–137 • (BIO) <br> Complete **Lesson 13 Worksheet 2** Pages 103–104 • (TG) | | | |
| | Day 78 | Read Laboratory 13, Cellular Respiration Pages 138–139 • (BIO) <br> Start Laboratory 13 Pages 105–106 • (TG) | | | |
| | Day 79 | Conclude Laboratory 13 and Prepare **Lesson 13 Lab Report** Page 107 • (TG) • Review Chapter 13 for quiz | | | |
| | Day 80 | Complete **Quiz 13** Page 261 • (TG) | | | |
| Week 8 | Day 81 | Begin Chapter 14, Chromosomes and Genes <br> Read Pages 140–143 • (BIO) <br> Complete **Lesson 14 Worksheet 1** Page 109 • (TG) | | | |
| | Day 82 | Finish reading Chapter 14 • Pages 144–145 • (BIO) • <br> Complete **Lesson 14 Worksheet 2** Page 110 • (TG) | | | |
| | Day 83 | Read Laboratory 14, Chromosomes and Genes Pages 146–149 • (BIO) • Start Laboratory 14 Pages 111–113 • (TG) | | | |
| | Day 84 | Conclude Laboratory 14 and Prepare **Lesson 14 Lab Report** Page 115 • (TG) • Review Chapter 14 for quiz | | | |
| | Day 85 | Complete **Quiz 14** Pages 263–264 • (TG) | | | |
| Week 9 | Day 86 | Begin Chapter 15, The Genetic Code • Read Pages 150–152 • (BIO) • Complete **Lesson 15 Worksheet 1** Page 117 • (TG) | | | |
| | Day 87 | Finish reading Chapter 15 • Pages 153–155 • (BIO) <br> Complete **Lesson 15 Worksheet 2** Page 118 • (TG) | | | |
| | Day 88 | Read Laboratory 15, The Genetic Code Pages 156–157 • (BIO) <br> Start Laboratory 15 Pages 119–120 • (TG) | | | |
| | Day 89 | Conclude Laboratory 15 and Prepare **Lesson 15 Lab Report** Page 121 • (TG) • Review Chapter 15 for quiz | | | |
| | Day 90 | Complete **Quiz 15** Page 265 • (TG) | | | |
| | | Mid-Term Grade | | | |

## Second Semester Suggested Daily Schedule

| Date | Day | Assignment | Due Date | ✓ | Grade |
|------|-----|------------|----------|---|-------|
| | | Second Semester-Third Quarter | | | |
| Week 1 | Day 91 | Begin Chapter 16, Expression of DNA — Transcription • Read Pages 158-161 to top paragraph • (BIO) Complete **Lesson 16 Worksheet 1** Page 123 • (TG) | | | |
| | Day 92 | Finish reading Chapter 16 • Pages 161 from top paragraph-163 • (BIO) • Complete **Lesson 16 Worksheet 2** Page 124 • (TG) | | | |
| | Day 93 | Read Laboratory 16, Transcription — mRNA Pages 164-167 • (BIO) • Start Laboratory 16 Pages 125-128 • (TG) | | | |
| | Day 94 | Conclude Laboratory 16 and Prepare **Lesson 16 Lab Report** Page 129 • (TG) • Review Chapter 16 for quiz | | | |
| | Day 95 | Complete **Quiz 16** Page 267 • (TG) | | | |
| Week 2 | Day 96 | Review Lesson 13 and Lesson 13 Quiz | | | |
| | Day 97 | Review Lesson 14 and Lesson 14 Quiz | | | |
| | Day 98 | Review Lesson 15 and Lesson 15 Quiz | | | |
| | Day 99 | Review Lesson 16 and Lesson 16 Quiz | | | |
| | Day 100 | Take **Exam 4 (Lessons 13–16)** Pages 301-302 • (TG) | | | |
| Week 3 | Day 101 | Begin Chapter 17, Expression of DNA — Translation • Read Pages 168-170 • (BIO) Complete **Lesson 17 Worksheet 1** Page 131 • (TG) | | | |
| | Day 102 | Finish reading Chapter 17 • Pages 171-173 • (BIO) Complete **Lesson 17 Worksheet 2** Page 132 • (TG) | | | |
| | Day 103 | Read Laboratory 17, Disruption of DNA Translation and Products of DNA Translation within Bacteria Cells Pages 174-177 • (BIO) • Start Laboratory 17 Pages 133-136 • (TG) | | | |
| | Day 104 | Conclude Laboratory 17 and Prepare **Lesson 17 Lab Report** Page 137 • (TG) • Review Chapter 17 for quiz | | | |
| | Day 105 | Complete **Quiz 17** Page 269 • (TG) | | | |
| Week 4 | Day 106 | Begin Chapter 18, Perpetuation of Life • Read Pages 178-181 • (BIO) • Complete **Lesson 18 Worksheet 1** Page 139 • (TG) | | | |
| | Day 107 | Finish reading Chapter 18 • Pages 182-185 • (BIO) Complete **Lesson 18 Worksheet 2** Page 140 • (TG) | | | |
| | Day 108 | Read Laboratory 18, Diverse Products of Protein Translation at Different Stages of Development Pages 186-189 • (BIO) Start Laboratory 18 Pages 141-143 • (TG) | | | |
| | Day 109 | Conclude Laboratory 18 and Prepare **Lesson 18 Lab Report** Page 145 • (TG) • Review Chapter 18 for quiz | | | |
| | Day 110 | Complete **Quiz 18** Page 271 • (TG) | | | |

| Date | Day | Assignment | Due Date | ✓ | Grade |
|---|---|---|---|---|---|
| Week 5 | Day 111 | Begin Chapter 19, Genetics Patterns I • Read Pages 190-193 • (BIO) • Complete **Lesson 19 Worksheet 1** Page 147 • (TG) | | | |
| | Day 112 | Finish reading Chapter 19 • Pages 194-195 • (BIO) Complete **Lesson 19 Worksheet 2** Page 148 • (TG) | | | |
| | Day 113 | Read Laboratory 19, Human Genetics Pages 196-199 • (BIO) Start Laboratory 19 Pages 149-152 • (TG) | | | |
| | Day 114 | Conclude Laboratory 19 and Prepare **Lesson 19 Lab Report** Pages 153-155 • (TG) • Review Chapter 19 for quiz | | | |
| | Day 115 | Complete **Quiz 19** Page 273 • (TG) | | | |
| Week 6 | Day 116 | Review Lesson 17 and Lesson 17 Quiz | | | |
| | Day 117 | Review Lesson 18 and Lesson 18 Quiz | | | |
| | Day 118 | Review Lesson 19 and Lesson 19 Quiz | | | |
| | Day 119 | Review Lessons 17–19 | | | |
| | Day 120 | Take **Exam 5 (Lessons 17–19)** Pages 303-304 • (TG) | | | |
| Week 7 | Day 121 | Begin Chapter 20, Genetics Patterns 2 • Read Pages 200-202 • (BIO) • Complete **Lesson 20 Worksheet 1** Page 157 • (TG) | | | |
| | Day 122 | Finish reading Chapter 20 • Pages 203-207 • (BIO) Complete **Lesson 20 Worksheet 2** Page 158 • (TG) | | | |
| | Day 123 | Read Laboratory 20, Dihybrid Test Cross with Corn Pages 208-211 • (BIO) • Start Laboratory 20 Pages 159-162 • (TG) | | | |
| | Day 124 | Conclude Laboratory 20 and Prepare **Lesson 20 Lab Report** Page 163 • (TG) • Review Chapter 20 for quiz | | | |
| | Day 125 | Complete **Quiz 20** Page 275 • (TG) | | | |
| Week 8 | Day 126 | Begin Chapter 21, Genetic Mutations and Variations • Read Pages 212-214 • (BIO) Complete **Lesson 21 Worksheet 1** Page 165 • (TG) | | | |
| | Day 127 | Finish reading Chapter 21 • Pages 214-219 • (BIO) Complete **Lesson 21 Worksheet 2** Page 166 • (TG) | | | |
| | Day 128 | Read Laboratory 21, Genetic Variation Pages 220-223 • (BIO) Start Laboratory 21 Pages 167-169 • (TG) | | | |
| | Day 129 | Conclude Laboratory 21 and Prepare **Lesson 21 Lab Report** Page 171 • (TG) • Review Chapter 21 for quiz | | | |
| | Day 130 | Complete **Quiz 21** Page 277 • (TG) | | | |
| Week 9 | Day 131 | Begin Chapter 22, Genomics • Read Pages 224-227 end of top paragraph • (BIO) Complete **Lesson 22 Worksheet 1** Page 173 • (TG) | | | |
| | Day 132 | Finish reading Chapter 22 • Pages 228 from first full paragraph-229 • (BIO) Complete **Lesson 22 Worksheet 2** Page 174 • (TG) | | | |
| | Day 133 | Read Laboratory 22, Plants Pages 230-233 • (BIO) Start Laboratory 22 Pages 175-177 • (TG) | | | |
| | Day 134 | Conclude Laboratory 22 and Prepare **Lesson 22 Lab Report** Page 179 • (TG) • Review Chapter 22 for quiz | | | |
| | Day 135 | Complete **Quiz 22** Page 279 • (TG) | | | |

| Date | Day | Assignment | Due Date | ✓ | Grade |
|------|-----|-----------|----------|---|-------|
| | | Second Semester-Fourth Quarter | | | |
| **Week 1** | Day 136 | Review Lesson 20 and Lesson 20 Quiz | | | |
| | Day 137 | Review Lesson 21 and Lesson 21 Quiz | | | |
| | Day 138 | Review Lesson 22 and Lesson 22 Quiz | | | |
| | Day 139 | Review Lessons 20–22 | | | |
| | Day 140 | Take **Exam 6 (Lessons 20–22)** Pages 305-306 • (TG) | | | |
| **Week 2** | Day 141 | Begin Chapter 23, Plant Taxonomy • Read Pages 234-238 • (BIO) Complete **Lesson 23 Worksheet 1** Page 181 • (TG) | | | |
| | Day 142 | Finish reading Chapter 23 • Pages 239-243 • (BIO) Complete **Lesson 23 Worksheet 2** Page 182 • (TG) | | | |
| | Day 143 | Read Laboratory 23, Protistans and Fungi Pages 244-247 • (BIO) Start Laboratory 23 Pages 183-186 • (TG) | | | |
| | Day 144 | Conclude Laboratory 23 and Prepare **Lesson 23 Lab Report** Page 187 • (TG) • Review Chapter 23 for quiz | | | |
| | Day 145 | Complete **Quiz 23** Page 281 • (TG) | | | |
| **Week 3** | Day 146 | Begin Chapter 24, Animal Taxonomy — Invertebrates • Read Pages 248-252 • (BIO) Complete **Lesson 24 Worksheet 1** Pages 189-191 • (TG) | | | |
| | Day 147 | Finish reading Chapter 24 • Pages 253-257 • (BIO) Complete **Lesson 24 Worksheet 2** Pages 191-192 • (TG) | | | |
| | Day 148 | Read Laboratory 24, Invertebrate Animals Pages 258-263 • (BIO) Start Laboratory 24 Pages 193-198 • (TG) | | | |
| | Day 149 | Conclude Laboratory 24 and Prepare **Lesson 24 Lab Report** Page 199 • (TG) • Review Chapter 24 for quiz | | | |
| | Day 150 | Complete **Quiz 24** Page 283 • (TG) | | | |
| **Week 4** | Day 151 | Begin Chapter 25 Animal Taxonomy — Vertebrates • Read Pages 264-270 • (BIO) Complete **Lesson 25 Worksheet 1** Pages 201-202 • (TG) | | | |
| | Day 152 | Finish reading Chapter 25 • Pages 271-277 • (BIO) Complete **Lesson 25 Worksheet 2** Pages 203-204 • (TG) | | | |
| | Day 153 | Read Laboratory 25, Vertebrate Animals Pages 278-283 • (BIO) Start Laboratory 25 Pages 205-208 • (TG) | | | |
| | Day 154 | Conclude Laboratory 25 and Prepare **Lesson 25 Lab Report** Page 209 • (TG) • Review Chapter 25 for quiz | | | |
| | Day 155 | Complete **Quiz 25** Page 285 • (TG) | | | |
| **Week 5** | Day 156 | Review Lesson 23 and Lesson 23 Quiz | | | |
| | Day 157 | Review Lesson 24 and Lesson 24 Quiz | | | |
| | Day 158 | Review Lesson 25 and Lesson 25 Quiz | | | |
| | Day 159 | Review Lessons 23–25 | | | |
| | Day 160 | Take **Exam 7 (Lessons 23–25)** Pages 307-308 • (TG) | | | |

| Date | Day | Assignment | Due Date | ✓ | Grade |
|---|---|---|---|---|---|
| Week 6 | Day 161 | Begin Chapter 26, Views of Biological Origins • Read Pages 284-188 • (BIO) Complete **Lesson 26 Worksheet 1** Page 211 • (TG) | | | |
| | Day 162 | Finish reading Chapter 26 • Pages 289-295 • (BIO) Complete **Lesson 26 Worksheet 2** Pages 213-214 • (TG) | | | |
| | Day 163 | Read Laboratory 26, Decomposition or Fossilization Pages 296-297 • (BIO) • Start Laboratory 26 Pages 215-217 • (TG) | | | |
| | Day 164 | Conclude Laboratory 26 and Prepare **Lesson 26 Lab Report** Page 219 • (TG) • Review Chapter 26 for quiz | | | |
| | Day 165 | Complete **Quiz 26** Page 287 • (TG) | | | |
| Week 7 | Day 166 | Begin Chapter 27, Evidences of Biological Origins • Read Pages 298-302 • (BIO) Complete **Lesson 27 Worksheet 1** Page 221 • (TG) | | | |
| | Day 167 | Finish reading Chapter 27 • Pages 303-305 • (BIO) Complete **Lesson 27 Worksheet 2** Page 222 • (TG) | | | |
| | Day 168 | Read Laboratory 27, Diversity Within Kinds of Creation Pages 306-307 • (BIO) • Start Laboratory 27 Pages 223-224 • (TG) | | | |
| | Day 169 | Conclude Laboratory 27 and Prepare **Lesson 27 Lab Report** Page 225 • (TG) • Review Chapter 27 for quiz | | | |
| | Day 170 | Complete **Quiz 27** Page 289 • (TG) | | | |
| Week 8 | Day 171 | Begin Chapter 28, Human Origins • Read Pages 308-311 • (BIO) Complete **Lesson 28 Worksheet 1** Pages 227-228 • (TG) | | | |
| | Day 172 | Finish reading Chapter 28 • Pages 312-315 • (BIO) Complete **Lesson 28 Worksheet 2** Pages 229-230 • (TG) | | | |
| | Day 173 | Read Laboratory 28, Human Origins Pages 316-319 • (BIO) Start Laboratory 28 Pages 231-232 • (TG) | | | |
| | Day 174 | Conclude Laboratory 28 and Prepare **Lesson 28 Lab Report** Page 233 • (TG) • Review Chapter 28 for quiz | | | |
| | Day 175 | Complete **Quiz 28** Page 291 • (TG) | | | |
| Week 9 | Day 176 | Review Lesson 26 and Lesson 26 Quiz | | | |
| | Day 177 | Review Lesson 27 and Lesson 27 Quiz | | | |
| | Day 178 | Review Lesson 28 and Lesson 28 Quiz | | | |
| | Day 179 | Review Lessons 26–28 | | | |
| | Day 180 | Take **Exam 8 (Lessons 26–28)** Pages 309-310 • (TG) | | | |
| | | Final Grade | | | |

# Worksheets

# and

# Laboratory Reports

## Matching

A. the unit membrane model          B. interspersed protein molecules

C. rubber wrapped around electrical wiring

1. _____ J.D. Robertson proposed

2. _____ The myelin sheath is like a blanket that insulates the nerve cell in the same way as

3. _____ In the fluid mosaic model of cell membranes, the cell membrane has two layers of phospholipid molecules with

## Fill in the Diagram

4.

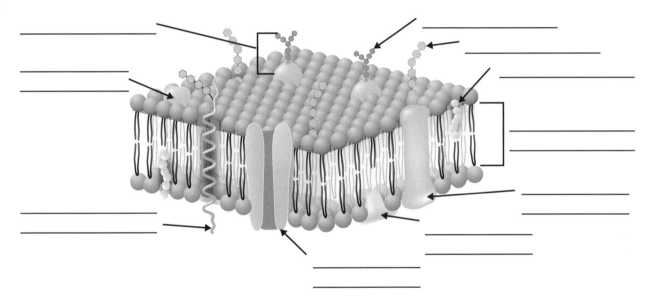

## Short Answer

5. The concept of membrane fluidity is that:

## Fill in the Blank

1. By saying that cell membranes are plastic, we mean that they have the ability to _____ _____.

2. The _____ is the control center of the cell.

3. DNA makes up the _____ that govern the inheritance of physical characteristics.

4. _____ contain DNA that has the genetic code to determine the structures of proteins within cells.

5. DNA and its associated proteins together is called _____.

6. The _____ in the nucleus of the cell contains RNA and protein.

7. The material inside of a cell between the outer membrane and the nuclear membrane is called _____.

8. Hammerling discovered that a new cytoplasm was governed by the _____ .

9. He used green algae called _____.

## Short Answer

10. Hammerling used two different species of algae, *A. mediterranea* and *A. crenulata*. If he used the base of *A. mediterranea* and the stalk of *A. crenulata*, a new cap would be regenerated that:

## Matching

      A. Nucleus          B. *A. mediterranea*          C. information

11. _____ When Hammerling cut off the cap of the newly formed algae, the new cap that grew matched the cap of this.

12. _____ Hans Spemann found that this part of the newt (salamander) embryo cells determined what the new salamander would look like.

13. _____ Spemann demonstrated after several cell divisions that the resulting cells have just as much genetic what as the original cell?

## Lab 6 — Cells

**Lab Notes:**

### REQUIRED MATERIALS

- ☐ Methylene blue stain (from the supply kit)
- ☐ Distilled water (available from grocery store)
- ☐ Microscope (from the supply kit)
- ☐ Microscope slide and cover slip (from the supply kit)
- ☐ Eyedropper (from the supply kit)
- ☐ Prepared slide *Spirogyra* (from the supply kit)
- ☐ Prepared slide *Paramecium (*fission) (from the supply kit)
- ☐ Prepared slide of bacteria cells (from the supply kit)
- ☐ Toothpicks

### INTRODUCTION

Mounting and staining cells is an important technique. Methylene blue is often used to stain cells such as those from your inner cheek. It is harmless and washes off easily. This lab provides experience observing mammalian cells (your cheek cells), algae cells (*Spirogyra*) that have external cell walls, protozoan cells (*Paramecium*), and bacteria cells (challenging because they are so small).

### PURPOSE

This laboratory exercise is designed to build upon the previous laboratory exercise by providing further experience in using the microscope, seeing the detailed structures of cells, and gaining experience in the preparation of wet mounts and observing prepared slides.

### PROCEDURE

1. Using the eyedropper, place a drop of distilled water in the middle of a clean microscope slide. With a clean toothpick, lightly scrape the inner lining of your cheek and smear it in the drop of water. Smear it around in the drop of water and add a drop of methylene blue. This will stain the cells making it easier to see them. Hold a cover slip over the stained cells and gently drop it onto the slide. It should spread out the liquid. Slides prepared this way are called wet mounts. Clean up around the edges of the coverslip by gently dabbing with a paper towel.

2. Place the slide onto the stage of the microscope. Focus on the slide by moving the lowest power objective lens until you can see the slide through the eyepiece. Use the smaller

fine focusing knob to bring the cheek cells better into focus. If there is too much material on the slide, you may need to redo the slide using less material from your cheeks. Try to identify the outer plasma membrane, cytoplasm, and nucleus of your cheek cells.

3. Rotate the turret and look at the slide through the medium power objective lens. The power of the eyepiece is 10x and the medium power objective lens is 10x, giving a total magnification of 100x. Describe what the cells look like. Draw what you see or take a picture with your phone. If you hold the phone up to the eyepiece you will be able to see the image on your phone. It is as if the phone is an additional lens between your eye and the eyepiece of the microscope. When you clearly see the image on your phone, take the picture. Can you see more detail?

4. Use the coarse focusing knob to raise the objective lens away from the slide. Rotate the turret so that you go back to the lowest power objective lens. Replace the slide with the prepared slide of *Spirogyra* (algae with cells connected in a long filament). Bring the lowest power objective lens toward the stage with the coarse focusing knob until you see the cells through the eyepiece.

5. *Spirogyra* is a single-celled organism. The cells are connected end to end. Algae are photosynthetic, meaning that they convert the energy of sunlight and $CO_2$ into glucose ($C_6H_{12}O_6$) and $O_2$. They are in the kingdom Protista (these are covered in chapters 23–25). This kingdom includes plant like algae, animal-like protozoa, and seaweed.

6. Rotate the turret so that you can look at the cells with the medium power objective lens. Look for the outer cell wall, nucleus, and chloroplasts (green objects within the cell). The chloroplasts contain the pigment chlorophyll that is used to capture energy from sunlight. Draw the *Spirogyra* using medium power or take a picture of it through the eyepiece with your phone.

7. Look at the *Spirogyra* slide under the highest power objective lens. How does it look different from the lower powers? Draw the *Spirogyra* under the high power or take a picture of it through the eyepiece with your phone.

8. Rotate the turret to the lowest power objective lens and move the objective lens away from the slide. Remove the *Spirogyra* slide and place the prepared *Paramecium* (fission) slide onto the stage. The *Paramecium* was dividing by fission (dividing in two) when the slide was prepared.

9. Look at the prepared slide of the *Paramecium* (fission) through the lowest, medium, and highest power objective lenses. Draw the *Paramecium* undergoing fission with the high power or take a picture of it through the eyepiece with your phone. Fission is asexual reproduction because the new cells have the same genetic material as the original cell. Identify the plasma membrane, nucleus, and gullet of the *Paramecium*. The gullet is where it takes in algae and other protozoa as food. Notice that the *Paramecium* does not have a cell wall like the *Spirogyra*.

   Like the *Spirogyra*, the *Paramecium* is also in the kingdom Protista. It is called a protozoan because it is animal-like in that it does not engage in photosynthesis. It has to eat its food, whereas the *Spirogyra* makes its food by photosynthesis. Notice that these names are in italics. That is because they are the genus names.

10. Examine the prepared slide of bacteria cells under the high power. Remember to start with the low power, then medium power, and afterwards the high power. Look for 3 types of bacteria cells — **coccus** (round), **rods,** and **spirals**. Describe and draw or take a picture of what you observe.

11. Answer the questions in this lab in your lab report with complete sentences.

**Laboratory Report** (20 points possible)

Description of cheek cells through the microscope

Drawing or picture of cheek cells through the microscope (plasma membrane, cytoplasm, and nucleus labeled)

Cheek cells under medium power

Cheek cells under high power

Drawing or picture of *Paramecium* under high power (label plasma membrane, nucleus [you may see more than one], and gullet)

Drawings and descriptions of observed bacteria

**Fill in the Blank**

1. By diffusion, molecules move from where they are _____ concentrated to where they are _____ concentrated.

2. Water is a _____ (other molecules will dissolve into it), and what dissolves into it is called the _____.

3. The process of _____ involves water diffusing across a cell membrane from a region of higher concentration to a region of lower concentration.

**Matching**

   A. Hyperosmotic      B. Isotonic      C. Preservative      D. Hyposmotic

4. _____ Physiological saline has 0.9% NaCl, the same as a living cell, and is called this (or isosmotic solution) because it is the same concentration on both sides of the cell membrane.

5. _____ The 0% NaCl concentration outside the cell is called hypotonic or this to the solution inside a cell.

6. _____ 10% NaCl is called hypertonic or this when contrasted with the solution inside a cell.

7. _____ Sugar acts as a what in jelly?

**Short Answer**

8. When water enters a cell by osmosis:

## Short Answer

1. It is active transport because:

## Fill in the Blank

2. Some of the proteins imbedded in the two phospholipid layers of the cell membrane are called _____ pumps.

3. An example is the _____-potassium exchange pumps in neurons (nerve cells).

4. The energy-releasing process is universal among _____ organisms.

5. When a nerve impulse passes down a neuron's membrane, protein gates allow _____ ions to rush into the cytoplasm.

6. The protein pump pumps _____ back out of the cell and _____ ions back into the cell.

## Matching

Energy release process in order of book:

A._____ → B._____ + C. _____ + D. _____

7. _____ ADP

8. _____ Energy

9. _____ Phosphate

10. _____ ATP

## Laboratory 7: Osmosis

**Lab Notes:**

### REQUIRED MATERIALS

☐ Fresh apple
☐ Fresh potato
☐ Distilled water (available from grocery store)
☐ NaCl (salt from grocery store)
☐ Knife to cut the potato and apple
☐ Teaspoon
☐ 100 ml beakers (3) (from the supply kit)
☐ Metric ruler (from the supply kit)
☐ 50 ml graduated cylinder (from the supply kit)
☐ Wax pencil (from the supply kit) or a non-permanent marker

### INTRODUCTION

Water diffuses into and out of cells through pores (called aquapores) in cell membranes. Diffusion occurs from a region where molecules are more concentrated to where they are less concentrated. The concentration of water in distilled water is 100% because there is nothing besides water. A 10% NaCl (salt) solution has a water concentration of 90%. The fluid inside of cells is 99.1% water, so if a cell is placed in a 10% NaCl solution, water will diffuse out of the cell more than will diffuse into the cell (from 99.1% to 90%) and the cell will shrivel up.

If cells are placed in distilled water (100% water), it will diffuse more into the cell with (99.1% water) than out of the cell and the cell will swell and perhaps burst.

### PURPOSE

To observe a practical example of osmosis.

### PROCEDURE

1. Label a beaker as distilled water, another beaker as 1.0% NaCl, and a third beaker as 10.0% NaCl.

2. Measure out a 50 ml (milliliter) portion of distilled water with a graduated cylinder and pour it into the first beaker.

3. Add 0.5 gram of NaCl to the 50 ml of distilled water in the second beaker — 0.5 gram is the amount on the tip of a table knife. This is accurate enough for this procedure. This beaker should be labeled as 1.0% NaCl.

4. Add 5 grams of NaCl to the water in the third beaker. One teaspoon of NaCl is 5 grams. This beaker should be labeled as 10.0% NaCl.

5. Cut 3 rectangles of potato and 3 rectangles of apple about 2 cm (centimeters) long, 1 cm tall, and 1 cm wide. Use the metric ruler from your supply kit.

6. Dry each of the potato and apple rectangles by blotting them with a paper towel.

7. Place a potato and apple rectangle in each of the 3 beakers.

8. After 30 minutes, remove the rectangles and place them on a paper towel. Blot each one with paper towel. Keep track of which pieces came from which beaker. Describe the appearance of each rectangle. Are they swollen, shriveled, or the same as before?

9. Write a complete sentence describing how the potato and apple pieces compare to their appearance before they were placed in the beakers.

10. **Osmosis** is the diffusion (movement) of water across a cell membrane from the side of the membrane where it is more concentrated to the other side of the membrane where it is less concentrated. Distilled water is 100% water. The beaker with 0.5 grams of NaCl in 50 ml of water is 1.0% NaCl and 99% water (100 − 1 = 99). The beaker with 5 grams of NaCl in 50 ml of water is 10% NaCl and 90% water (100 − 10 = 90).

   A. For the first beaker, the potato and apple pieces with 99% water are place in distilled water (100%). Would you expect the water to move more into or more out of the potato and apple cells? From your observations what do you think happened? If more water moved into the potato and apple cells, they would swell up. If more water moved out of the potato and apple cells, they would shrink. If the same amount of water entered the cells as came out of them, they would stay pretty much the same. Did it happen the way you thought it would?
   B. For the second beaker, the potato and apple pieces with 99% water are placed in 1% NaCl (99% water). The fluid within the cells is 99.1% water to begin with. Would you expect the water to move more into or out of the potato and apple cells? From your observations, what do you think happened? Did it happen the way you thought it would?
   C. For the third beaker, the potato and apple pieces with 99% water are placed in 10% NaCl (90% water). Would you expect the water to move more into or more out of the potato and apple cells? From your observations what do you think happened? Did it happen the way you thought it would?

11. Answer the questions in this lab in your lab report with complete sentences.

**Laboratory Report** (20 points possible)

Appearance of potato rectangle before the procedure

Appearance of apple rectangle before the procedure

Observations of potato rectangle after being in distilled water

Observations of apple rectangle after being in distilled water

Observations of potato rectangle after being in 1% NaCl (0.9% is inside the cells)

Observations of apple rectangle after being in 1% NaCl

Observations of potato rectangle after being in 10% NaCl

Observations of apple rectangle after being in 10% NaCl

Expectations of water moving into or out of the cells in distilled water

Observations of water movement into or out of the cells in distilled water

Expectations of water moving into or out of the cells in 1% NaCl water

Observations of water moving into or out of the cells in 1% NaCl water

Expectations of water moving into or out of the cells in 10% NaCl water

Observations of water moving into or out of the cells in 10% NaCl water

Evaluation of expectations of the cells in distilled water

Evaluation of expectations of the cells in 1% NaCl

Evaluation of expectations of the cells in 10% NaCl

## Fill in the Blank

1. _____ are barrel-shaped organelles in _____ cells.

2. There are usually _____ of these structures in each cell, positioned at right angles to each other.

3. In cell division, they move to _____ ends of a cell.

4. Plants have _____ that move chromosomes to opposite ends of a cell.

5. Centrioles also form long whip-like tails called _____ found on some cells.

6. They also form small hair-like structures called _____ that act like oars in a boat, propelling a cell through water.

7. _____ are important for releasing energy from food molecules and oxygen.

8. The inner folds of these structures are called _____.

9. Contrary to creation, some evolutionists think that mitochondria were formed when _____ invaded cells.

10. This is called the _____ theory.

11. When an organism is well fed, mitochondria _____ and become more _____.

12. _____ are produced in the nucleolus.

## Fill in the Diagram

13.

## Matching

| A. Smooth | B. Reticulum | C. Toxic | D. Circulatory | E. Rough |
| --- | --- | --- | --- | --- |

14. _____ Ribosomes attach to an extensive system of flattened membranes called the endoplasmic what?

15. _____ The space between these large, sheet-like organelles acts like the _____ system of a cell.

16. _____ When ribosomes are attached to these structures, these structures are called _____ ER.

17. _____ When ribosomes are not attached to these structures, they are called _____ ER.

18. _____ Smooth ER with enzymes attached break down _____ molecules in blood.

## Short Answer

1.  Write out what lysomes release:

## Matching

A.  Chloroplasts          B.  Grana          C.  Water          D.  Plastids

2.  _____ Plant cells have small organelles called this in their cytoplasm that animal cells do not have.

3.  _____ Photosynthesis occurs in what?

4.  _____ They have stacked membranes called this.

5.  _____ Grana contain chlorophyll, carotene, and enzymes necessary to convert carbon dioxide and this into glucose (sugar) and oxygen, using energy from sunlight.

## Fill in the Blank

6.  Plant cells store starch as an energy reserve in _____.

7.  In some plant cells, the _____ contain red and yellow pigments.

8.  Plants have _____ _____ that cover their cells and support and protect them.

9.  These structures have the polysaccharide _____ that most animals cannot digest.

10. Plant cells sometimes have a large _____ _____ that contains mostly water and acts like the kidney of plant cells.

11. Some protozoa have _____ _____ that help them expel excess water.

## Laboratory 8: Cell Structures

**Lab Notes:**

### REQUIRED MATERIALS

- ☐ Microscope (from the supply kit)
- ☐ Prepared microscope *Paramecium* slide (from the supply kit)
- ☐ Prepared microscope *Amoeba* slide (from the supply kit)
- ☐ Prepared microscope human blood smear slide (from the supply kit)
- ☐ Prepared microscope frog blood smear slide (from the supply kit)

### INTRODUCTION

In this laboratory exercise you will gain further experience to refine your skill in using a microscope. The *Paramecium* and *Amoeba* (also spelled *Ameba*) are single-celled animal-like cells under the broad category Protozoa in the kingdom Protista. *Paramecium* are common in bodies of fresh water in ponds, streams, and lakes. They feed upon single-celled algae (such as *Spirogyra* that you studied before) and other protozoa. In aquatic food webs, they are eaten by larger protozoa and multicellular animals. *Amoebas* also feed upon other protozoa and algae. They move and feed by changing their shape. They extend their pseudopodia (false feet) and the rest of the cell catches up with them. They feed by wrapping their pseudopodia around their prey and releasing enzymes from lysosomes that digest the prey. This is called amoeboid motion. The study of *Amoebas* has greatly helped in the understanding of our white blood cells that behave in the same way. In His grace, God has given us other organisms to study to further our understanding of ourselves. For example, God created the eyes of squid to be very much like mammalian eyes, which has helped us understand our own eyes.

Erythrocytes (red blood cells) are much smaller than protozoa. They are responsible for transporting $O_2$ from our lungs to our cells. Because of their small size, they will challenge your microscope skills.

### PURPOSE

This lab will further develop your microscope skills by observing and describing *Paramecium*, an *Amoeba*, and red blood cells and some of their structures.

## PROCEDURE

1.  Place the prepared *Paramecium* conjugation slide onto the stage of the microscope and examine it under low power. Conjugation is the process whereby two *Paramecia* swap nuclei giving each other new genetic material. Draw or take a picture of the *Paramecium* and describe what you see with complete sentences. Look for the structures in the following diagram.

2.  Examine the *Paramecium* under the medium and high powers. Draw or take a picture and describe what you see with complete sentences. Look carefully to see if you can see the larger macronucleus and the smaller micronucleus. The macronucleus is more involved with the living processes of the *Paramecium* and the micronucleus is more involved with reproduction. They are swapped in conjugation. Paramecia are called ciliates because they are protozoa with cilia (small hair-like structures sticking out from their outer surface). Cilia are very small, so you may not be able to see them.

3.  Place the prepared *Amoeba* slide onto the stage of the microscope and examine it under low power. Draw or take a picture of the *Amoeba* and describe what you see with complete sentences. Look for the structures in the following diagram.

4.  Examine the *Amoeba* under the medium and high powers. Draw or take a picture and describe what you see with complete sentences.

5.  Place the prepared human blood smear slide onto the stage of the microscope and examine it under low power. Red blood cells look like very tiny almost transparent doughnut-shaped structures. You will also see a few larger stained white blood cells. Draw or take a picture of the blood sample and describe what you see with complete sentences. Look for the structures in the following diagram.

6.  Examine the human blood smear under the medium and high powers. Hemoglobin in red blood cells contain iron combined with oxygen (the same as rust), which gives blood its red color. They are very small and thin, so they look almost transparent under the microscope. Draw or take a picture and describe what you see with complete sentences. Under the higher power, the red blood cells appear as pale doughnut-shaped ghosts with their middle slightly filled in.

7.  Examine the prepared frog blood smear slide under the high power of the microscope. Describe how it is different from human blood. Propose an idea as to why God created frog blood different from human blood. Hint — frogs get a lot of additional oxygen through their skin and we do not.

8.  Complete the lab report as instructed in the teacher's guide.

**Laboratory Report** (20 points possible)

Drawing or picture and description of *Paramecium* with low power

Drawing or picture and description of *Paramecium* with medium power

Drawing or picture and description of *Paramecium* with high power

Can you see the macronucleus and smaller micronucleus?

Drawing or picture and description of *Amoeba* with low power

Drawing or picture and description of *Amoeba* with medium power

Drawing or picture and description of *Amoeba* with high power

Drawing or picture and description of human red blood cells with high power

Drawing or picture and description of frog blood cells with high power

## Short Answer

1. Virchow's Cell Law states that:

## Fill in the Blank

2. An egg cell combined with the DNA from a sperm cell produces a single cell called a

   _____.

3. The division of the cell in question 2 is called _____ because the resulting cells are

   _____.

4. Before a cell divides, its _____ is _____.

5. The two strands of DNA _____ (come apart), and next to each a

   _____ strand is formed.

6. Now there are two sets of two strands of DNA where each consists of an _____ strand and a

   _____ strand.

7. The life cycle of a cell is called the _____ _____.

8. _____ is what happens to a cell between cell divisions.

9. The process in question 8 has _____ parts called _____, _____, and _____.

10. In the G1 phase, the cell _____ to normal size.

11. In the S (synthesis) phase, _____ is replicated and most of the _____ are
    duplicated.

12. In the G2 phase, the cell grows more and _____ to _____.

13. During _____, the cell nucleus divides.

14. During _____, the cytoplasm divides.

## Matching

   A. Diploid          B. Asexually          C. Centromere          D. Sister

15. _____ Cell division enables some organisms to reproduce this (producing another organism from one
    organism).

16. _____ Cells with two of each chromosome are called this.

17. _____ When a chromosome replicates in the S phase, it produces 2 pairs of what kind of Chromatids?

18. _____ The two objects formed in question 17 are connected by a protein structure called this.

## Matching

A.  Apart          B.  Prophase          C.  Coiled          D.  Condensed

1.  _____  The membrane around the nucleus comes apart in this.

2.  _____  In the phase of mitosis called prophase, the chromosomes become this.

3.  _____  The chromosomes become _____, thicker, and shorter; otherwise, they would be strung out too thin.

4.  _____  In prophase, the membrane around the nucleus comes _____.

## Fill in the Blank

5.  _____ _____ attach to the centromeres and pull apart the chromatids.

6.  At the end of this phase, there are _____ of each chromosome.

7.  The next phase of mitosis is _____, where the chromosomes are pulled to the _____ of the cell.

8.  During the third phase of mitosis, called _____, chromosomes can be seen being pulled toward the _____ ends of the cell.

9.  During the fourth phase of mitosis, called _____, the chromosomes can be seen at the _____ ends of the cell with the new _____ reforming around them.

10. There are now _____ nuclei with _____ of each chromosome in each.

11. In animal cells, a _____ _____ forms during cytokinesis, separating the two _____ of the cell.

12. In plant cells, a _____ _____ forms between the new cells and gradually separates them.

13. An "n" stands for the _____ of _____ in a cell.

14. In mitosis, 2n → _____ → _____ x _____.

## Short Answer

15.  Psalm 51:5 says:

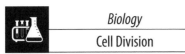
## Laboratory 9: Cell Division

**Lab Notes:**

### REQUIRED MATERIALS

- ☐ Microscope (from the supply kit)
- ☐ Microscope Prepared Slide Onion Root Tip Mitosis (from the supply kit)
- ☐ Prepared Slide of *Ascaris* (roundworm) Mitosis (from the supply kit)

### INTRODUCTION

Plant roots grow at their tips where there is a cap that leads the way to penetrate soil. The region where cell division occurs in the root tip is called meristematic tissue. Not all cells in an organism divide. When a cell matures and takes on specialized functions (called differentiation) only part of the DNA is expressed, giving them their specialized functions — such as nerve cells or muscle cells. Within most tissues, some of the cells do not differentiate and retain the ability to divide. These cells are called **stem cells** because they can still divide, producing new cells to replace worn-out cells and new cells for growth. Within the meristematic tissue, just inside the cap of the root tip of a growing onion plant, the cells are dividing, producing new tissue.

When you examine the prepared slide with the chromosomes stained in the meristematic tissue of a root tip, you should be able to see cells in the various stages of cell division.

### PURPOSE

In this lab exercise you will gain further experience using a microscope to observe fine detail at high magnification. You will see the chromosomes of the cells at the various stages of mitosis and cytokinesis. This helps to connect physical reality with the description of mitosis in the text. To some extent, we are all visual learners and you will understand the process of cell division much better if you have seen it for yourself. If someone told a group of people about elephants that they have never seen, they would each come away with different images of an elephant. Hopefully, you will not see an elephant with the microscope.

### PROCEDURE

1. Place the prepared slide of the onion root tip onto the stage of the microscope and examine it under the low power objective lens.

2. Carefully follow the procedure you learned in the earlier lab exercise going from low to medium to high power. You should be able to see individual cells in the tip of the onion root. Move the slide around slowly so that you can see several of the cells.

3. As you observe the slide in the following steps, look for and identify each stage of mitosis on the slide, show it to your teacher and describe what you are looking at and what takes place in that stage. Your teacher will use how well you find the stages of mitosis and your comments in grading this lab exercise.

4. Go to the medium power and look at the slide. Go to the higher power to more clearly see the stages of mitosis on the slide.

5. Look for a cell in **prophase** where you can see the nucleus with the stained chromosomes. Look around for a cell without a nuclear membrane but with the chromosomes that look just like those in the cell with a nuclear membrane. It is still in prophase.

6. Look for a cell in **metaphase** where the chromosomes appear lined up in a line down the middle of the cell.

7. Look for a cell in **anaphase** where the chromosomes are moving to the opposite ends of the cell. The cell should appear to be a bit stretched out as well.

8. Look for a cell in **telophase** where the chromosomes are definitely at opposite ends of the cell and you may be able to see new nuclear membranes forming around each group of chromosomes. Before cytokinesis occurs, it will look like the cell has 2 nuclei.

9. Since these are plant cells, a new **cell plate** forms down the middle separating the new daughter cells. The cell plate will form cell walls between the new cells. See if you can see 2 daughter cells stuck together with a thick cell plate between them.

10. Look at the prepared slide of the *Ascaris* cross section showing cells in cell division. This is from an animal (roundworm) rather than a plant, so you will not see a cell wall. You should see the new cells separating by being pinched off in the middle. How many of the stages of mitosis can you find? List and briefly describe them.

11. Your teacher will grade your lab report according to the instructions in the teacher's guide.

**Laboratory Report** (20 points possible)

Point out and describe cells of the onion root tip in each of the following stages of mitosis to your teacher.

Prophase

Metaphase

Anaphase

Telophase

Point out and describe the stages of cell division that you can find on the *Ascaris* slide to your teacher.

You will be graded based upon accuracy and thoroughness.

## Short Answer

1. Species are:

## Fill in the Blank

2. The study of the interrelationships of organisms and their environment is called _____.

3. In the Bible, we are told that God created all organisms in different _____.

4. A group of organisms in the same species living in the same region is called a _____.

## Matching

A. Ecosystem      B. Gene pool      C. Microevolution
D. Macroevolution      E. Community

5. _____ All of the genetic variations within this group are called a:

6. _____ All of the living organisms living in the same region are a:

7. _____ All of the living organisms in the same region and everything non-living in that region make up the:

8. _____ Small amount of change within a kind is sometimes referred to as:

9. _____ Darwin said many small changes for vast periods of time would result in large changes called:

## Fill in the Blank

1. The area occupied by a population is called its _____.

2. Everything an organism needs from day to day is its _____.

3. The number of mice of the same species that occupy the same area is an example of

   _____ _____.

4. The limit as to the number of mice that the area can support is its _____

   _____.

## Matching

                A. Primary succession        B. Secondary succession        C. Succession

5. _____ The gradual change in the species of plants in the same ecosystem over time is called:

6. _____ When an island is formed by an underwater volcano, and when plants and animals move in and become established on the island, the process is called:

7. _____ After a fire goes through an area, grasses grow first and fill in the area. When seeds from shrubs and trees germinate and grow in the shelter of the grasses, the process is called:

## Short Answer

8. A climax community is reached:

9. The text example of this is:

## Laboratory 10: Ecosystems

**Lab Notes:**

### REQUIRED MATERIALS

- ☐ A notepad (something to write on) and a pencil
- ☐ 1 square foot of earth — **Optional**
- ☐ 1 square foot of screening or chicken wire — **Optional**
- ☐ Small bone (chicken wing or drumstick) — **Optional**

### INTRODUCTION

An ecosystem is a region where several populations can be found. A population is a group of organisms of the same species (similar to each other and can mate and reproduce). The ecosystem description includes all of the populations found there (the community), the ground and air (non-living parts) and the climate of the area. Usually, you will only see part of an ecosystem at one time. Within an ecosystem you will see many microclimates. For example, the temperature down in a squirrel hole can be at least 10°F cooler than the ground surface on a warm day. The cool shade under a tree is a microclimate as is the surface of a hot rock in the sun.

### PURPOSE

The concept of an ecosystem and how to study it will be developed in this exercise.

### PROCEDURE

1. Go to a place outdoors of your choosing. This could even be in your back or front yard. Describe where you are — include the city, state, country, open meadow, lawn area, forest, etc.

2. Use these guidelines to prepare this lab report:

   A. Use complete sentences in all of your descriptions.
   B. What is the date and time that you are making your observations?
   C. Describe the weather of your location. This refers to the conditions at the time that you are writing this report. Climate refers more to long-range conditions.

3. Describe 4 or 5 populations that you can see in this ecosystem. This refers to different plants and/or animals. You can use common names or physical descriptions. Do not worry about looking up genus and species names at this time.

4. Describe relationships that you can observe between the organisms you described — such as predator, prey, providing shelter (like a tree for a bird), etc. It may take some time observing to notice interactions between them. Take the time to enjoy and learn from this exercise. You may want to photograph or sketch some of them. This is the largest part of your report.

5. Do any of the organisms (animals or plants) appear to be overly crowded (nearing carrying capacity), about right, or very sparse (very few in the area)?

6. Describe what you think this area will be like 6 months in the future. Include the weather, non-living and living parts of the ecosystem.

**Optional Enrichment Exercise**

7. This is provided if you are up for a challenge and have the time. This may sound weird (because it is) but this step is preparation for an optional portion of laboratory 26. From a meal (or a neighbor's meal) take a chicken leg (or similar bone) with a little bit of meat left on it. You are going to prepare a fossil. Find a small dirt area outside about a foot square. If you do not have an area like this, check with someone close by who will let you use their property. Dig down about 2 inches. Place the bone in the middle of the area and sprinkle a little wet dirt on it. Take some screen material or chicken wire and form a "roof" over the dug-out dirt area. This is to prevent critters from messing with it. Each day, sprinkle a little more wet dirt over it until the depression in the ground is covered. Check on it periodically to be sure that nothing messes it up. Let it stay there until you come to laboratory 26. Keep a journal of weekly observations. Describe how and where you set up your "fossil." Do you have to reinforce the wire cage to protect it? What is the weather like? Is it wet or dry? If you cannot be there for a given week, have someone else check on it for you. Be sure to write down the time and date of every observation. You can do it more than once a week, but do it at least once a week. This journal will form part of your optional lab report for laboratory 26. Be sure to always use complete sentences, and your descriptions have to be clear enough for someone else to understand them without you being there to explain them.

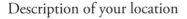
**Laboratory Report** (20 points possible)

Description of your location

Date and time

Weather

Description of 4 or 5 populations that you observe

Observations of the relationships between the populations

Observations of any populations that appear to be overly crowded, about right, or too sparse

What you think the area will be like in 6 months (include weather, non-living and living parts of this ecosystem)

**Optional Enrichment Exercise**

Keep weekly observations starting next week. Follow the directions in step 7 for making your observations.

Week 1

Week 2

Week 3

Week 4

Week 5

Week 6

Week 7

## Fill in the Blank

1. The geographical regions of North America are divided into _____, depending upon the climate, plants, and soil.

2. Grasslands are exposed to extreme _____ between seasons.

3. The great "dustbowl" was caused by increased erosion by _____ and _____.

4. Much of the free-roaming herbivore mammals, such as _____, in these areas were replaced by _____.

5. _____ biomes have extremely low rainfall totals.

6. The rainstorms coming their way many times lose their moisture as they pass over the _____.

7. _____ formation may be a major source of water for desert plants and animals.

## Short Answer

8. Describe metabolic water and give an example:

## Matching

A. Coniferous        B. Mammals        C. Deciduous        D. Shade

9. _____ In these forests, trees lose their leaves every fall.

10. _____ These trees usually produce this, which keeps many plants from growing under them.

11. _____ These trees produce cones.

12. _____ You will find few of these in the most-dense forests.

## Short Answer

1. Based on the author insights, why should tropical rain forests be protected?

## Fill in the Blank

2. These are the most _____ and _____ biomes on earth.

3. _____ takes place rapidly on the forest floor because of the higher temperatures.

4. _____ plays a significant role in the germination of many seed types in this region.

5. The _____ biome is found along the west coast of the United States.

6. This area has many _____ bushes that do well in drier conditions.

7. The trees and shrubs in this region tend to be less than _____ _____ tall.

8. Many of the plants require _____ to reproduce.

9. The climate of chaparral is called a _____ climate.

10. _____ is the study of freshwater environments.

11. The organisms that live on or in the bottom of aquatic ecosystems are called _____.

12. The term _____ includes aquatic life forms that can move considerable distances in the direction of their choosing.

13. _____ waters are poor in minerals and other nutrients, and therefore poor in animal diversity.

14. When water accumulates excess nutrients, it is called _____.

15. _____ and _____ are some of the richest aquatic habitats for photosynthesis.

## Matching

A. Death        B. 71%        C. 33%        D. Genetic

16. _____ The growth of land animals is limited by these constraints and gravity.

17. _____ Many marine organisms continue to grow until their _____.

18. _____ Marine aquatic ecosystems cover about this percent of earth's surface.

19. _____ The total marine biome contributes about this percent of the earth's entire photosynthetic activity.

## Laboratory 11: Biomes

### REQUIRED MATERIALS

☐ Notepad (something to write on or computer) and a pencil

### INTRODUCTION

A biome is a large area that has several features of climate, soil, and plant and animal life in common. Review chapter 11 to identify your biome. If you know several traits of a biome in general, it helps you understand several features around you that may not be readily apparent.

### PURPOSE

This laboratory is to help you understand the biome where you live and to give you experience in gaining information and describing the larger area in which you live. For example, fires are common to the chaparral biome, so you should be careful and aware of fire if you live there.

### PROCEDURE

1. Review chapter 11 and identify the biome in which you live. Write out in complete sentences the general traits of your biome. You should go beyond the chapter and consult other sources. Cite the references for any sources that you use. You should include enough information in your reference so that you could go right back to it at a later time.

2. Write out characteristics of your region of the biome. This can come from your personal observations, a local nature center or similar facility, and observations from traveling short distances around your home.

3. Describe how the region around your home matches up to the general description of your biome. Describe any local variations that may be different from the general traits of your biome. Use things such as climate, soil, different animals (such as birds and insects), different plants, and any variations such as bodies of water, large open meadows or mountains or hills.

4. How does your part of the biome vary with the seasons — such as summer and winter?

**Lab Notes:**

| *Biology* | Pages 118–119 | Day 64 | Lesson 11 | Name |
|-----------|---------------|--------|-----------|------|
| Biomes | | | Laboratory Report | |

**Laboratory Report** (20 points possible)

Identify your biome

Observations of your biome

Observations of how your biome resembles or differs from the general biome described in your text

How does your part of the biome vary over the seasons?

## Short Answer

1.  Briefly describe how plants are producers and autotrophs:

## Matching

A.  Ultraviolet      B.  Red      C.  Pigments      D.  Heterotrophs      E.  Infrared

2.  _____ Animals (and humans) are called this because they cannot make their own food.

3.  _____ In the visible light spectrum, this color has the least energy.

4.  _____ This light has less energy than red light.

5.  _____ This light has more energy than violet light.

6.  _____ Molecules that can absorb and reflect light energy are called this.

## Fill in the Blank

7.  _____ and other pigments trap light energy in plants and algae.

8.  Englemann demonstrated that algae were absorbing the _____, _____, _____, and _____ colors of light and reflecting _____ light.

9.  Plant cells are said to show _____ _____, meaning that if they did not have all their parts they could not function.

10. _____ use energy from sulfur compounds instead of light to produce glucose and oxygen.

11. The _____ states that systems left to themselves tend to come apart and lose energy.

12. _____ and _____ are essential processes in photosynthesis and respiration.

13. _____ is the removal of electrons and the energy stored in the chemical bonds.

14. _____ is the reverse — it is adding electrons.

15. Photosynthesis is the _____ of carbon.

16. Respiration is _____ that releases energy.

## Fill in the Blank

1.  In the Light Reactions, _____, _____, and _____ are produced.

2.  _____ provides energy for the Dark Reactions.

3.  _____ provides hydrogen and electrons for the Dark Reactions.

4.  _____ plants live under moderate climate conditions.

5.  Trees are _____ plants.

## Fill in the Diagram

6.              Light Reactions Produces                      Dark Reactions Produces

## Matching

              A.  $CO_2$          B.  Closed          C.  Night          D.  $C_4$

7.  _____ These plants live under hotter, drier conditions.

8.  _____ To prevent water loss, the stomata of plants in hot, dry conditions remain like this during part of the day.

9.  _____ This is stored in 4 carbon molecules in the bundle sheath cells of plants in hot, dry conditions so that it can be released when needed.

10.  _____ CAM plants only open their stomata during this time.

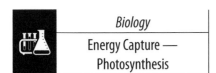

| | *Biology* | Pages 128–129 | Day 68 | Lesson 12 Laboratory | Name |
|---|---|---|---|---|---|
| | Energy Capture — Photosynthesis | | | | |

## Laboratory 12: Photosynthesis

**Lab Notes:**

### REQUIRED MATERIALS

☐ 5 2-inch-long segments of water plant (from pet shop or aquarium section of other stores, strands of algae found along a pond or stream will also work)
☐ 5 large test tubes (22 ml) (from the supply kit)
☐ 1 beaker (100 ml) (from the supply kit)
☐ Bright light source
☐ Distilled water
☐ Green, yellow, and red food coloring
☐ Phenol red pH indicator (from the supply kit)
☐ 1 drinking straw
☐ Timer — perhaps your smart phone
☐ Wax pencil (from the supply kit)

### INTRODUCTION

Photosynthesis in plants and algae converts $CO_2$ and $H_2O$ into glucose ($C_6H_{12}O_6$) and oxygen ($O_2$), using energy from sunlight or other bright light source. The photosynthetic pigments in plants absorb particular wavelengths of light. When a plant is green, it is absorbing colors other than green and reflecting green light back to you.

### PURPOSE

In this procedure, you are to test the effectiveness of different colors of light in photosynthesis. At first, all wavelengths of visible light are exposed to plant material and then the colors green, yellow, and red will be eliminated to see their effectiveness. By blowing air into the water with a drinking straw, you are supplying $CO_2$ which is converted into $H_2CO_3$ ($CO_2 + H_2O \longleftrightarrow H_2CO_3$). The solution bathing the plant material will have $H_2CO_3$ (carbonic acid) as a source of $CO_2$. The phenol red pH indicator in the solution will turn yellow as you blow $CO_2$ into it. As the $CO_2$ is used, more $H_2CO_3$ breaks down to replace the $CO_2$ until there is no more $H_2CO_3$ left. The faster photosynthesis occurs, the less time it takes to use up the $CO_2$ in the solution. When the plant uses up the $CO_2$, the $H_2CO_3$ decreases until the solution is no longer acid, turning the phenol red indicator pink. You will determine how long it takes for the solution containing the plant to turn from yellow to pink. The color of light that is most effectively used in photosynthesis will require the least amount of time to use up the $CO_2$.

## PROCEDURE

1. Fill 5 test tubes almost to the top with distilled water. Add 1 drop of phenol red pH indicator to each test tube. Place about a 2-inch segment of water plant into each test tube. Number the tubes from 1 to 5.

2. Place test tube #5 aside to be observed later.

3. With a straw, blow air into test tube #1 until the solution turns yellow, indicating that it is acidic. Place it in a beaker of clear water with a bright light shining on the beaker. Determine how long it takes to turn from yellow to pink. Record this time.

   As $H_2CO_4 \rightarrow CO_2 + H_2$) and $CO_2$ is used by the water plant for photosynthesis, the solution becomes more basic (less acid) and the phenol red goes from yellow to pink.

4. Blow air into test tube #2 until the solution turns yellow. Place it in a beaker with green water (made green with a drop of green food coloring) with a bright light shining on the beaker. Determine how long it takes to turn pink. Record this time.

5. Blow air into test tube #3 until the solution turns yellow. Place it in a beaker with yellow water with a bright light shining on the beaker. Determine how long it takes to turn pink. Record this time.

6. Blow air into test tube #4 until the solution turns yellow. Place it in a beaker with red water with a bright light shining on the beaker. Determine how long it takes to turn pink. Record this time.

7. Check test tube #5. Has it turned pink? If it did, did it take longer than test tubes #1 through 4?

8. Report your results in a table like the one below.

| Test tube number | 1 | 2 | 3 | 4 | 5 |
|---|---|---|---|---|---|
| Time it takes to turn pink | | | | | |

9. Which color(s) is most effective in water plant photosynthesis? This would be the one that turned the phenol red pink in less time. The bright light has to shine through the water in the beaker to reach the test tube. For example, for test tube #2, the green water reflects green light back to you and allows red, orange, yellow, blue, and violet light to reach the test tube. Likewise, for test tube #3, the yellow light is reflected and does not reach the test tube.

10. Test tube #5 is called a control. Why would that be? Hint — it is not modified like the others.

11. In your report, list your results in a table and answer the questions with complete sentences.

## Laboratory Report (20 points possible)

Time it takes the solution to turn pink

Test Tube #1 _____

Test Tube #2 _____

Test Tube #3 _____

Test Tube #4 _____

Test Tube #5 _____

What color is most effective for water plant photosynthesis?

How did you decide which color was most effective?

Why is test tube #5 called the control?

What happened with test tube #5?

## Short Answer

1. Write out the overall equation for respiration.

## Fill in the Blank

2. Cellular respiration occurs in the _____ of plant and animal cells.

3. Cellular respiration occurs in three stages called _____, _____
   _____ _____, and _____ _____
   _____.

4. _____ does not require oxygen to occur.

5. Because this stage does not require oxygen, it is said to be _____.

6. The _____ _____ cycle and _____
   _____ _____ do require oxygen and are said to be
   _____.

7. Using oxygen to add phosphate to ADP, and thus forming ATP, is called _____
   _____.

8. NAD molecules remove _____ atoms and their electrons and takes them to the Electron Transport
   System.

9. If there is not sufficient oxygen supplied, an _____ _____ occurs and
   the pyruvic acid is converted into _____ _____, which causes muscle
   cramps.

## Matching

      A. Cytoplasm      B. Matrix      C. Glucose      D. Cristae

10. _____ The lactic acid is carried to the liver where it is converted back into:

11. _____ Glycolysis occurs in the fluid portion of the cell's:

12. _____ The inner membrane of the mitochondrion is called the:

13. _____ The fluid within this inner membrane is called the:

## Fill in the Blank

1. The third carbon atom of pyruvic acid is given off as a molecule of _____ that is eventually exhaled by the lungs.

2. The 2 carbon molecule _____ combines with the 4 carbon molecule _____ _____ to form the 6 carbon molecule _____ _____.

3. The _____ carbon molecule of citric acid becomes a _____ carbon molecule of alpha-ketoglutaric acid, which becomes the _____ carbon molecule of succinic acid.

4. Throughout the citric acid cycle (also called _____ Cycle), NAD removes _____ atoms and takes them to the Electron Transport System.

5. _____ and _____ carry _____ ions to the matrix of the mitochondria.

6. H+ ions build up in the intermembrane space and pass back into the matrix through the _____ _____ complex that produces ATP molecules by the process called _____.

## Matching

          A.  2               B.  38             C.  O=

7. _____ H+ ions combine with what kind of negative-charged ions to produce metabolic water?

8. _____ In glycolysis, how many ATP are gained from each glucose molecule?

9. _____ In the Electron Transport System, there is a total yield of how many ATP molecules from each glucose molecule?

**Fill in the Diagram**

10.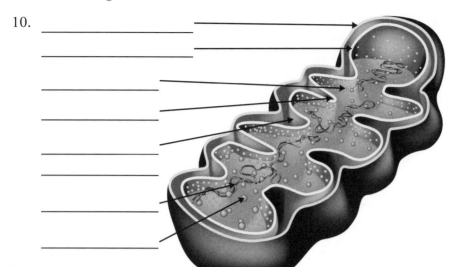

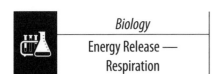

| Biology | | | | Name |
|---|---|---|---|---|
| Energy Release — Respiration | Pages 146–149 | Day 78 | Lesson 13 Laboratory | |

## Laboratory 13: Cellular Respiration

**Lab Notes:**

### REQUIRED MATERIALS

- ☐ Distilled water
- ☐ 3 large test tubes (from the supply kit)
- ☐ 3 small test tubes (from the supply kit)
- ☐ 3 100 ml beakers (from the supply kit) of warm water
- ☐ Yeast (1 package of baker's yeast)
- ☐ ¼ teaspoon measuring spoon
- ☐ ½ cup measuring cup
- ☐ Glucose (also called dextrose) (from the supply kit)
- ☐ Sucrose (table sugar)
- ☐ Ruler (to measure mm, millimeters) (from the supply kit)
- ☐ Wax pencil (from the supply kit)

### INTRODUCTION

Yeast (a single-celled fungus) goes through glycolysis but not Kreb's Cycle or the Electron Transport System. Since glycolysis does not require $O_2$, they are anaerobic. This also means that they do not generate as much ATP as aerobic organisms. At the end of glycolysis, they convert pyruvic acid into ethanol (ethyl alcohol $C_2H_5OH$). This process is called fermentation. The overall equation for their respiration is . . .

$$C_6H_{12}O_6 + 2\ ADP \rightarrow 2\ CO_2 + 2\ C_2H_5OH + 2\ ATP$$
(glucose)

### PURPOSE

This procedure is designed to measure and contrast the $CO_2$ production of yeast starting from glucose ($C_6H_{12}O_6$) and sucrose ($C_{12}H_{22}O_{11}$). Sucrose is a glucose molecule bonded to a fructose molecule (sugar found in fruit).

### PROCEDURE

1. Add ¼ teaspoon of yeast to ½ cup of distilled water and mix well.

2. Add ¼ teaspoon of glucose (dextrose) and ¼ teaspoon of yeast to ½ cup of distilled water and mix well.

3. Add ¼ teaspoon of sucrose and ¼ teaspoon of yeast to ½ cup of distilled water and mix well.

4. Mark 3 small test tubes ⅔ up from the bottom and number the tubes from 1 to 3.

5. Fill small test tube #1 to the ⅔ mark with the yeast solution.

6. Fill small test tube #2 to the ⅔ mark with glucose yeast solution.

7. Fill small test tube #3 to the ⅔ mark with sucrose yeast solution.

8. Add distilled water to each of the test tubes from steps 5–7 to the top.

9. Place a large test tube upside down over each small test tube.

10. With your finger, push the small test tube up against the bottom of the large test tube and quickly invert the test tubes so that the small test tube is upside down in the large test tube. Be sure to keep the top of the small test tube up against the inner bottom of the larger test tube. Try to keep the amount of air going into the small test tube to a minimum. You might want to practice this a few times. Do this for all 3 sets of test tubes.

11. Make a mark on each set of test tubes at the level of air in the smaller test tube.

12. Place the 3 sets of test tubes in the beakers of warm water. As the yeast converts sugar to $CO_2$, the gas in the small test tubes will increase. Leave them in the warm water for about an hour — or less if the tube is filling rapidly with $CO_2$. Leave the 3 sets of tubes in the warm water for the same amount of time.

13. Mark each tube at the final gas ($CO_2$) level. Measure the increase in $CO_2$ in each tube with a ruler as mm (millimeters). Which produced the most $CO_2$ — the one with glucose, sucrose, or no sugar? The least amount of $CO_2$? Try to explain your results, keeping in mind that glycolysis begins with glucose. Test tube #1 is called a control. Why is that? Record your results in your report using a table like the following example. What would it mean if test tube #1 produced more $CO_2$ than test tubes #2 and #3? Hopefully, it does not.

| Test tube number | 1 | 2 | 3 |
|---|---|---|---|
| mm of $CO_2$ produced | | | |

**Laboratory Report** (20 points possible)

Level of $CO_2$

Test Tube #1 _____

Test Tube #2 _____

Test Tube #3 _____

Which test tube produced the most $CO_2$?

Which test tube produced the least $CO_2$?

Explain the results.

Why is test tube #1 called the control?

What would it mean if test tube #1 produced more $CO_2$ than test tubes #2 and #3?

**Short Answer**

1. The term chromosome came about because:

**Matching**

        A. 23        B. Deoxyribose        C. Diploid        D. 2

2. _____ We have eukaryotic cells, most of which are:

3. _____ Humans have this many pairs of chromosomes:

4. _____ DNA has a double helix structure, meaning that it is composed of this many strands coiled around each other:

5. _____ The sugar in DNA is:

**Fill in the Blank**

6. A sugar, phosphate, and base together are called a _____.

7. For every guanine in DNA, there is a _____ in the opposite strand.

8. For every cytosine in DNA, there is a _____ in the opposite strand.

9. For every adenine in DNA, there is a _____ in the opposite strand.

10. For every thymine in DNA, there is an _____ in the opposite strand.

11. _____ bonds hold the opposing strands of DNA together.

12. In 1958, Meselson and Stahl demonstrated that DNA is replicated in a _____ fashion.

13. This means that the new DNA has a _____ strand and an _____ strand.

## Fill in the Blank

1. E. coli was used to demonstrate the information from the strands, extracting its DNA both before and after it replicated with $^{15}N$ _____.

2. This resulted in new cells with DNA that had a lighter strand with _____ and a heavier strand with _____.

3. In a centrifuge, the DNA from the new E. coli cells moved through the cesium chloride _____ and _____ than the original DNA.

4. _____, also called a _____ division, results in cells where there is only one of each chromosome.

5. The resulting cells with one of each chromosome are called _____.

## Matching

A. Regulator      B. Chromosome      C. Gene      D. Sperm

6. _____ Children can inherit DNA from the egg and the:

7. _____ This is a sequence of DNA that codes for a protein chain:

8. _____ When a cellular process, like the Krebs cycle, has several enzymes that function in sequence, the genes that code for the enzymes are usually in the same order in a:

9. _____ This kind of gene determines whether a sequence of genes should be turned on or off:

## Short Answer

10. When a cellular process, like Krebs cycle, has several enzymes that function in sequence:

## Laboratory 14: Chromosomes and Genes

**Lab Notes:**

### REQUIRED MATERIALS

- ☐ Microscope (from the supply kit)
- ☐ Microscope Prepared Slide Onion Root Tip Mitosis (from the supply kit)
- ☐ Microscope Prepared Slide Roundworm *Ascaris* (from the supply kit)
- ☐ 3 lengths of thick string 30 inches long
- ☐ 81 paper clips
- ☐ 14 strips of red colored paper ½ inch x 1 inch
- ☐ 16 strips of blue colored paper ½ inch x 1½ inch
- ☐ 14 strips of green colored paper ½ inch x 1 inch
- ☐ 16 strips of yellow colored paper ½ inch x 1½ inch
- ☐ 16 strips of white paper ½ inch x 1 inch
- ☐ Other colors of paper can be used, as long as you have 5 different colors

### INTRODUCTION

Chromosomes are long lengths of double-stranded DNA with attached proteins. Genes are sections of DNA that contain the information to direct the formation of a protein (or protein subunit). There are many genes connected end to end making up a chromosome. As humans, we have at least 25,000 genes on 23 chromosomes. DNA is structured like a twisted ladder with 2 alternating supports on the outside and rungs in between. The outside supports are made up of alternating 5 carbon sugar molecules (deoxyribose) and phosphates ($PO_4{}^{-3}$). The rungs of the ladder are the bases (adenine, thymine, guanine, and cytosine) that are attached to the deoxyribose molecules.

Along the length of a DNA strand, every group of 3 bases is called a **codon**. Each codon codes for an amino acid. In the process called *transcription* a single-stranded length of RNA is made by copying part of one of the strands of the DNA. This RNA strand called messenger RNA (mRNA) leaves the nucleus and goes out into the cytoplasm where it is "read" to construct a protein which is a length of amino acids.

### PURPOSE

You are constructing a model to visualize the structure of a chromosome. When you read a description like that of the structure of a chromosome, the imagination is a tricky thing and can come up with different ideas. This exercise is designed to help you have an accurate image of the chromosome structure. This is essential to understand the following chapters.

## PROCEDURE

1. Look at the prepared slide with the metaphase stage of mitosis of the onion root tip. Start with the low power objective lens and shift to the medium and then to the highest power objective lens. The darkly stained "stringy"-looking structures that you see in the middle of the cell are chromosomes. These are composed of extremely long strands of DNA. Look for chromosomes on the *Ascaris* slide as well. Point out the chromosomes on the onion root tip slide and the *Ascaris* slide to your teacher.

2. Put a capital letter A (adenine) on each of the blue (1½ inch long) strips of paper. Set two of these aside for next week.

3. Put a capital letter G (guanine) on each of the yellow (1½-inch-long) strips of paper. Set two of these aside for next week.

4. Put a capital letter T (thymine) on each of the red (1-inch-long) strips of paper.

5. Put a capital letter C (cytosine) on each of the green (1-inch-long) strips of paper.

6. Put a capital letter U (uracil) on each of the white (1-inch-long) strips of paper. Set two of these aside for next week.

7. Attach the paper clips to each length of string at 1-inch intervals with 27 on each length of string.

8. Lay 2 of the lengths of string on a flat surface parallel to each other with their paper clips facing each other. To each paper clip on one string connect one of the paper strips. Mix up the colors.

9. Connect paper strips to each of the paper clips on the parallel string so that A is opposite T and C is opposite G between the 2 strings.

10. Draw the "double stranded DNA" that you constructed.

11. On your diagram, draw a line after every 9 bases. This should give you 3 sections with 9 bases on each string.

12. Draw a line after each group of 3 bases within the groups of 9 bases.

13. Each group of 9 bases represents a gene that codes for 3 amino acids.

14. Each group of 3 bases represents a codon that codes for an amino acid.

15. In your DNA, each group of 3 bases is a codon and each group of 9 bases is a gene that codes for a protein consisting of 3 amino acids. Real proteins may have over 100,000 amino acids — that would take a lot of string and paper clips.

16. Separate the strings and place the third string with paper clips between the two original strips with its paper clips facing the left-hand string.

17. Attach strips of paper to the paper clips on the third string so that A (third string) faces T (first string); U (third string) faces A (first string); G (third string) faces C (first string) and C (third string) faces G (first string).

18. Notice that the third string has U instead of T. RNA has uracil instead of thymine.

19. Slide the third string out from between the first and second strings and bring the first and second strings back together again.

20. The third string represents a single stranded messenger RNA (m-RNA) molecule that leaves the nucleus of the cell to direct the amino acid sequence of a new protein formed in the cytoplasm of the cell. This process is called **transcription,** which comes from the word *transcribe*, which means to write out a copy. The original double-stranded DNA (represented by the first and second strings) remains protected in the nucleus of the cell.

21. Your lab report is oral, consisting of pointing out and describing the chromosomes on the onion root tip slide and your model of DNA and the process of transcription.

22. Save the 3 strands (strings with paper clips and bases attached to them) that you made for the next lab.

**Laboratory Report** (20 points possible)

Point out and describe the chromosomes on the onion root tip slide and *Ascaris* slide to your teacher.

Drawing of your double-stranded DNA with a line drawn every 9 bases and every 3 bases.

Describe how you made the mRNA from the DNA.

Draw the mRNA that you constructed. Describe your DNA model, how the mRNA was constructed, and the mRNA to your teacher.

## Short Answer

1. What did Irwin Chargaff find about DNA in 1950?

## Fill in the Blank

2. An mRNA strand is made from _____ of the DNA strands.

3. Each section of 3 bases in DNA is called a _____.

4. Each of these is translated into an _____ acid in the sequence of a

   _____.

## Matching

        A. Leucine         B. Proline         C. Glutamic

5. _____ Using the genetic code chart on page 152 of the student book, which amino acid is coded for by CCA in the mRNA?

6. _____ Using the genetic code chart on page 152 of the student book, which amino acid is coded for by UUG in the mRNA?

7. _____ Using the genetic code chart on page 152 of the student book, which amino acid is coded for by GAG in the mRNA?

## Fill in the Blank

1. The codon UGA in the mRNA indicates the _____ of an amino acid chain.

2. The codon AUG in the mRNA indicates the _____ of an amino acid chain.

3. The codon UAA in the mRNA indicates the _____ of an amino acid chain.

4. The genetic code is called _____ because more than one codon can stand for the same amino acid.

5. This demonstrates God's _____ in His design for life in a fallen world.

6. A form of a gene that produces a normal protein that is in the greatest number in the same species is the _____ _____.

7. Variations of the same gene are called _____.

8. A mutated gene that is rare in an organism is a _____ gene.

9. The percent of an allele in a population is the _____ _____.

## Matching

|  A. Frequency |  B. Genotype |  C. Phenotype |

10. _____ The two copies of each gene for an individual is the:

11. _____ The physical expression of the genes of an individual is the:

12. _____ Usually, the term "evolution" refers to the change in the allele:

## Short Answer

13. Why did the author suggest it was so important that in the DNA research it became evident that the code was universal and applied to all organisms — single-celled and multicellular?

*Biology*

The Genetic Code

Pages 156–157

Day 88

Lesson 15
Laboratory

Name

## Laboratory 15: The Genetic Code

### REQUIRED MATERIALS

- ☐ The 3 strands of string with the bases attached from laboratory 14
- ☐ 6 paper clips
- ☐ The 6 paper strips you set aside in laboratory 14 (AUGUGA) to attach to the paper clips for the third strand (mRNA)

### INTRODUCTION

The genetic code consists of the bases (codons) in the DNA strand that are copied to make a strand of mRNA. Each group of 3 bases directs the placement of an amino acid in a protein. When a protein is made, the first step is to link the amino acids together in the proper sequence. This directs the remaining steps in forming a functioning protein. The chart in this chapter indicates the amino acid placed in sequence for each group of 3 bases in the mRNA strand. From the smallest prokaryotes to multicellular eukaryotes, the same genetic code is used. It is possible that different combinations of bases could code different amino acids in different organisms, especially if they evolved over vast periods of time. The genetic code also demonstrates the grace of God by its redundancy. Many times, if one base is changed into another, the same amino acid is placed in a protein. The key to survival and quality of life is in the functioning of the proteins. So, if you had a base substitution and the appropriate protein was produced anyway, you would not know the difference.

### PURPOSE

The purpose of this laboratory exercise is to visualize the application of the genetic code. This exercise also provides important experience in keeping details straight. It is not always as easy as it may seem, but can be very important in many experiences in life. This skill is transferable to many different tasks that you may encounter. No matter what fields of study you might pursue in days ahead, you will have to deal with your genetic history, and need to make decisions based upon your genetic makeup and background. This is shown in the ways available to research your genetic history and apply it to health issues that may come up. This was not a major consideration just one generation ago.

**Lab Notes:**

## PROCEDURE

1.  Take the DNA strand produced in laboratory exercise 14 that was used to make the mRNA strand and list the bases in sequence. An example is . . .

Adenine, Cytosine, Thymine, Guanine, Thymine, Cytosine, etc.

From here on, the bases will be indicated as A, G, T, and C. For RNA, U will be used for uracil (RNA has uracil in place of thymine).

2.  Parallel to the list of bases for the DNA strand, list the bases in the third strand (mRNA). You have to have a start codon (AUG for methionine) and a stop codon (UGA). You need to add these to the beginning and end of your third strand because they were not placed there in laboratory #14. An example is . . .

| | |
|---|---|
| DNA | T A C A C G G T C A A T C G G T A G A C T |
| RNA | A U G U G C C A G U U A G C C A U C U G A |

The AUG (met which indicated the start of the sequence) was added to the beginning of the RNA sequence and UGA (stop) was added to the end of the sequence.

3.  Separate the bases of the mRNA into codons. An example is . . .

| | |
|---|---|
| RNA | A U G | U G C | C A G | U U A | G C C | A U C | U G A |

4.  Use the chart in chapter 15 to identify which amino acids it will place in the protein sequence. An example is . . .

| | | | | | | | |
|---|---|---|---|---|---|---|---|
| RNA | A U G | U G C | C A G | U U A | G C C | A U C | U G A |
| Protein | met (start) | cys | gln | leu | ala | ile | stop |

Notice that the RNA is like the DNA except that RNA has G where DNA has C; RNA has C where DNA has G; RNA has A where DNA has T and RNA has U where DNA has A. Your lab report consists of the results from steps 1 through 4 and the answers to the following questions.

5.  Did you make sure to add start and stop codons to your DNA and RNA sequences? Did you have any redundancies in your sequence? If you did, how would you recognize them? Show how a change in one base in your DNA sequence could have coded for a different amino acid. In the above example, if the GTC (third codon in the DNA) were GTG instead, the RNA would have CAC instead of CAG for its third codon. This would give pro (proline) for the second amino acid instead of gln (glutamine). Depending upon where this amino acid lies in the protein, it could have major consequences in the functioning of the protein.

**Laboratory Report** (20 points possible)

Describe your results for . . .

Step 1

Step 2

Step 3

Step 4

Answer the questions in step 5.

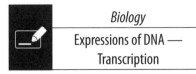
## Fill in the Blank

1. Proteins are long chains of _____ _____.

2. Amino acids are attached to each other to form protein chains while attached to _____.

3. The formation of mRNA from particular genes in the chromosomes is called _____.

4. To _____ means to rewrite, and DNA and mRNA are in the same language of a

   _____ _____.

## Matching

A. Nucleus          B. Introns          C. Exons          D. Nuclear

5. _____ The mRNA goes through the pores in this membrane and out to the cytoplasm:

6. _____ DNA is protected inside this part of a cell:

7. _____ Sections of mRNA called this are removed from mRNA:

8. _____ These remaining sections of RNA are spliced (connected) together to form shorter lengths of mRNA:

## Fill in the Diagram

9. Label the part of the diagram of the beginning process of protein synthesis:

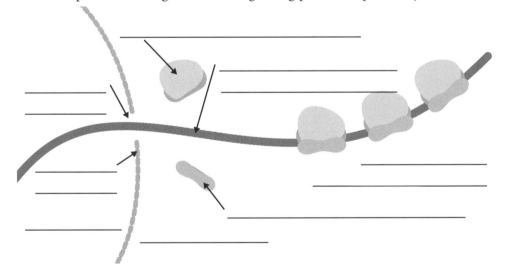

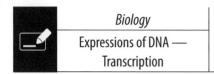
## Fill in the Blank

1. The word _____ in Genesis is different from the word _____ used today.

2. Some different _____ can interbreed to form hybrids.

3. When different species result from removing different introns, this is not _____, because it is necessary that all of the DNA was there from the beginning.

4. We know that the removal of introns is a controlled, designed (not random) process because it is controlled by _____, which are proteins coded for by DNA.

## Matching

       A. Protein        B. Alternative        C. Frameshift        D. Protein

5. _____ This occurs if the enzyme cuts the original mRNA in the wrong place.

6. _____ This totally distorts the resulting _____.

7. _____ This kind of splicing occurs when different combinations of exons are spliced together.

8. _____ This process means that the same DNA in the same organism can produce slightly different _____ molecules at different seasons of the year.

## Short Answer

9. When you listen to lectures or read materials presenting an evolutionary explanation remember this:

| | Biology | Pages 164–167 | Day 93 | Lesson 16 Laboratory | Name |
|---|---|---|---|---|---|
| | Expressions of DNA — Transcription | | | | |

## Laboratory 16: Transcription — mRNA Processing

**Lab Notes:**

### REQUIRED MATERIALS

☐ Paper and pencil

### INTRODUCTION

After mRNA is formed by transcription from a strand of DNA, it has to be modified before it can be used to code for a functioning protein. Parts of the mRNA (called **introns**) are removed and the remaining parts (called **exons**) are spliced together to code for a protein in the process of translation. Several similar but different proteins can be formed from the same mRNA by removing different introns. Different forms of the same protein work well under different conditions. For example, one form of a protein may function well at 70°F, another at 90°F, and another at 45°F. This is God's grace coming through creation, allowing organisms to survive as conditions change. This may seem like a complicated way to do it, but complexity is never an obstacle to God. As well, it seems that the more complex something is there are more ways for it to go wrong. But this is another way that God triumphs. He has produced many safeguards and repair enzymes to sustain life.

Jesus told His disciples that ". . . the very hairs of your head are all numbered" (Luke 12:7; NKJV). He actively cares for the smallest details — even to be sure that the important proteins in our bodies can function well when circumstances change.

God cares for all living things because He is their Creator and Sustainer. When God addressed Job about His greatness He said:

> *Can you hunt the prey for the lion, or satisfy the appetite of the young lions, when they crouch in their dens, or lurk in their lairs in wait? Who provides food for the raven, when its young ones cry to God, and wander about for lack of food? (Job 38:39–41; NKJV).*

God sustains life in its smallest details. When you consider what could go wrong, you realize that He is there all of the time maintaining His created work.

*All things were created through Him and for Him. And He is before all things, and in Him all things consist (Colossians 1:16–17; NKJV).*

You will see in this lab exercise that if an error occurred called a frameshift, the results would be disastrous. But by His sustaining power, thousands of proteins are produced every minute in the cells throughout your body and they work without frameshifts.

## PURPOSE

The purpose of this lab exercise is to visualize:

1.  The process that occurs in processing mRNA — removing introns and splicing the remaining exons

2.  The process of a frameshift and its effects

## PROCEDURE

1.  This sentence shows how introns and exons work. Each word has 3 letters because the words (codons) of DNA and RNA consist of 3 nucleotides.

    **THE BIG TAN DOG RAN TOO FAR FOR
    THE RED CAT WHO WAS NOT FAR.**

    If the "introns" removed are "big, cat, and not"; the remaining exons linked together are

    **THE TAN DOG RAN TOO FAR
    FOR THE RED WHO WAS FAR.**

    This is similar to the original sentence but still different.

    Take the original sentence and select a different set of "introns." It can include some of those in the example and some others. Then splice your "exons" together and form a new sentence. Notice that your new sentence is like the original but different.

    Do it again with another set of introns. This is going to take some of your creative juices because there are only so many possibilities that you can work with.

2. Consider the following sequence of nucleotides in an original mRNA molecule.

**U U U G A C C U C G C A G G C U C G A A G A C G U G C**

If this were the final form of the mRNA, the protein produced would have the amino acid sequence of . . .

**PHE, ASP, LEU, ALA, GLY, SER, LYS, THR, CYS**
(using the chart in chapter 15)

Notice that U (uracil) is used and not T (thymine) because we are dealing with RNA.

If the introns are GAC and UCG, the remaining exons spliced together produce

**U U U C U C G C A G G C A A G A C G U G C.**

This modified sequence of nucleotides codes for a protein with the amino acid sequence . . .

**PHE, LEU, ALA, GLY, LYS, THR, CYS**

Notice that it is like the original except the amino acids coded for by the introns are missing.

Select a different set of introns and splice your exons together and write out the amino acid sequence for which it would code. Again, notice that it is like the original but with some amino acids missing. You can see how this shows how 2 different forms of the same protein can be formed.

This is not an example of evolution because it starts with the unprocessed RNA already there plus all of the necessary enzymes (also proteins) already coded for and present for the process to begin with. This requires an initial creation with all of its complexity to begin with.

In the future you will interact with many well-educated and highly intelligent people who will claim that this complexity came about from simpler forms over long periods of time. Remember, they are not allowing for a Creator God in their thinking, so they have no choice but to assume that it came about by itself. It is not the data that convinces them, but their own personal choice. This should not compel you to doubt God and His Word. Do show respect and care, because you do not know if they may accept Christ and understand the true origin of the mechanisms of transcription.

3. Take the sequence of nucleotides from your modified mRNA. Take out 1 nucleotide (a C, G, U, or A) and write out the list of codons that you have left and determine the amino acid sequence that this altered mRNA would produce. For example, if you take the example from step 2 . . .

UUUCUCGCAGGCAAGACGUGC

and remove the first C, you get . . .

UUUUCGCAGGCAAGACGUGC

Instead of producing

PHE, LEU, ALA, GLY, LYS, THR, CYS,

you would get . . .

PHE, SER, GLN, ALA, ARG, ARG,

and the remaining GC would not be used.

This is the effect of a frameshift. The sequence is the frame and it shifted over by 1 nucleotide. This would completely destroy the protein. By God's grace, the enzymes that control the process prevent such a thing from happening.

4. In your lab report include all of the work from steps 1–3.

**Laboratory Report** (20 points possible)

Step 1

Step 2

Step 3

## Short Answer

1. Translation is:

## Fill in the Blank

2. When mRNA leaves the nucleus of a cell, _____ attach to it.

3. Many ribosomes attached to the same mRNA form a _____.

4. Amino acids first attach to _____ _____ molecules (tRNA), which are partially _____ stranded.

5. The codon AGU codes for the amino acid serine. The codon _____ will be on the bottom of the tRNA (transfer RNA) that attaches to the amino acid serine.

## Matching

A. Enzymes        B. Amino        C. Activator

6. _____ The match between the codons of mRNA and tRNA indicate the appropriate _____ acid that should be placed in the protein chain.

7. _____ An _____ enzyme attaches an amino acid to its appropriate tRNA.

8. _____ There has to be at least 20 different activator _____ because there are 20 different amino acids.

## Matching

A.  P          B.  Irreducible          C.  Codon          D.  A

1. _____ This interaction between nucleic acids and proteins is one of the aspects of _____ complexity in life.

2. _____ The _____ is named after "amino acid."

3. _____ The first amino acid-tRNA combination moves to the _____ site on the ribosome where the polypeptide chain grows.

4. _____ Each time a new amino acid tRNA combination attaches to the ribosome, the ribosome slides down to the next _____ on the mRNA.

## Fill in the Blank

5. A _____ code is designed by God's grace because if part of the code is altered, the ribosome will still stop instead of making a useless or destructive _____.

6. When there is no longer a need for more of a protein, no more _____ for that protein will be made. This is another terrific example of _____.

## Short Answer

7. Transcription is:

8. Translation means:

## Laboratory 17: Diverse Products of Protein Translation within an Organism

### REQUIRED MATERIALS

- ☐ Microscope (from the supply kit)
- ☐ Prepared slide of Muscle Types (from the supply kit)
- ☐ Prepared slide of Motor Neurons (from the supply kit)
- ☐ Antibiotic discs (ampicillin, erythromycin, neomycin, and penicillin) **Optional — for next week**
- ☐ *E. coli* live cultures — **Optional**
- ☐ Prepared Tryptic Soy agar media plates — **Optional**
- ☐ Sterile swab applicators — **Optional**
- ☐ Bleach — **Optional**
- ☐ Wax pencil (from the supply kit) — **Optional**
- ☐ 100 ml beaker about half full of bleach (from the supply kit) — **Optional**

### INTRODUCTION

Stem cells can produce different specialized cells depending on the products of protein translation, which depends upon the mRNA produced in transcription. In this lab exercise, striated (skeletal) muscle, smooth muscle, and cardiac muscle are examined. Stem cells in the right place develop into the proper muscle type to fulfill important roles. Striated muscles are used for voluntary functions such as walking, talking, or just sitting down. They are long chains of many cells connected end to end. They are called striated because they have dark vertical bands when viewed under the microscope. Smooth muscles appear as individual spindle-shaped cells. They are involved in involuntary functions such as moving food through the digestive tract, constricting veins to control blood flow, and squeezing glands to release their products into the blood stream. Cardiac muscle cells are connected at different points and are connected so as to form branches in different directions. This helps move the contractions faster through the heart. Impulses, similar to nerve impulses, travel from cell to cell through the heart to produce coordinated contractions.

### Optional Enrichment Exercise

Many times the role of an antibiotic is to prevent the production or functioning of important proteins. When the antibiotic neomycin is applied to a culture of *E. coli* (*Escherichia coli*) bacteria, it binds to the ribosomes and

**Lab Notes:**

prevents mRNA from being read to produce key proteins. Ampicillin penetrates the outer membranes of bacteria cells and inhibits the enzyme transpeptidase that is needed to form new bacteria cell walls. You are also looking at the effects of penicillin and erythromycin upon bacteria cells.

Most antibiotics act by preventing the reproduction of new bacteria cells. Since bacteria cells do not live very long, an important way to eliminate them is to keep them from reproducing.

A method of developing antibiotic resistance in bacteria is for a mutation to occur that damages a protein to which the antibiotic attaches. When altered, the protein cannot attach to the antibiotic molecule, which disables the antibiotic. When reference is made to bacteria evolving antibiotic resistance, the DNA is altered (damaged) rather than enhanced. This does not improve the genetic make-up of a cell even though it has an advantage whenever a particular antibiotic is introduced. When that antibiotic is not around, the DNA is still damaged and there may be other harmful consequences.

This optional exercise will continue into next week. The first thing to do is to grow a healthy culture of *E. coli* bacteria on agar with nutrients to enhance their growth. Next week you should be able to add paper discs that contain the antibiotics ampicillin, erythromycin, neomycin, and penicillin and observe their effects. This strain of *E. coli* bacteria is harmless. Another strain of *E. coli* is in your small intestine digesting the food you eat. This is sort of like ordering prechewed steak at a restaurant. (Yuck!)

In this optional exercise you are observing the consequences of disrupting the process of DNA translation and the alteration of proteins produced by DNA translation.

## PURPOSE

The purpose of this exercise is to observe how different products formed by different mRNA molecules result in different cell types being produced. This is seen in the three muscle types (striated, smooth, and cardiac) produced in a vertebrate body. Different stem cells form motor neurons. They have very different anatomy and function than muscle cells.

An additional optional enrichment exercise is provided to observe the consequences of disrupting DNA translation (neomycin preventing the attaching of mRNA to ribosomes) and the disruption of the functions of proteins produced by DNA translation. This will depend upon your resources and time.

## PROCEDURE

1. Examine the prepared slide of muscle types under high power. First, look at the striated muscle cells. Notice the nucleus and striations (dark bands). When the muscle contracts, the striations come closer together. The striations are perpendicular to the length of the muscle. Muscle cells are linked end to end very closely, so it is hard to tell where one ends and the other begins. A good indicator is that each cell had one nucleus before they joined together. You can count the cells by counting the nuclei. Point out the striated muscle cells and their striations and nuclei to your teacher. Describe their function.

2. Examine the smooth muscle cells under high power. There are no striations in smooth muscle. The cells are usually very close together and spindle shaped, but if you look close you should be able to tell that they are separate cells. Their nuclei are longer than in striated muscle. Point out the cells and their nuclei to your teacher. Describe their function.

3. Examine the cardiac muscle cells under high power. These cells are more branched and separated by intercalated discs. The intercalated discs are separations between the cells with pores that allow materials and impulses to be passed between them. This ensures that muscle contractions will rapidly travel from one cell to another. Their nuclei look very different than the other muscle types. If you look carefully, you may be able to see how the cells connect in branching patterns. Point out the cells and their nuclei to your teacher. Describe their function.

4. Examine the motor neurons under high power. Even though they are very different from muscle cells, they have the same DNA as muscle cells. An organism begins when the nucleus (with 1 of each chromosome) from a sperm cell combines with the nucleus of an egg cell (which also has 1 of each chromosome) producing a one-celled zygote (that has 2 of each chromosome). The zygote goes through many cycles of mitosis where all of the cells produced have identical DNA (2 of each chromosome). The resulting cells still become all of the different cells in the body. The cells differ by which parts of the DNA is used to produce different proteins. Some become muscle cells, and some become neurons.

   As you look at a neuron, notice that it has many branches sticking out that look like tree roots. The central part of the cell is the **cell body** that contains the **nucleus**. The shorter branches are the **dendrites** that receive neural impulses from other neurons. The longer branches are the **axons** that carry impulses away to other neurons or muscle cells. Find these structures on a neuron and point them out to your teacher. Describe the functions of the parts of a neuron and the overall function of a motor neuron. The word *motor* refers to motion, so a motor neuron carries impulses to muscle cells, causing them to contract.

**Optional Enrichment Exercise**

1. Take 2 of the prepared petri dishes and mark the tops with a wax pencil as shown in this diagram.

2. Mark the 4 sections of each of the 2 petri dishes as shown below with the letters A, E, N, and P standing for the antibiotics ampicillin, erythromycin, neomycin, and penicillin.

3. Remove a sterile swab from its container, being very careful not to touch the cotton portion. Carefully remove the cap from the tube containing the *E. coli* bacteria. Do not lay the cap down anywhere. Stick the sterile swab into the tube containing the *E. coli* and without digging into the agar, rub the cotton swab onto the bacteria contaminating the cotton swab. Then replace the cap onto the tube containing the *E. coli*.

4. Carefully partially lift the top of each petri dish (one at a time) and rub the contaminated cotton swab onto the agar in back and forth motions without digging into the agar. Then place the top back onto the petri dish so as to prevent anything else from getting onto the agar.

5. Do the same thing to the second petri dish.

6. Place both prepared petri dishes in a warm (not hot) place where they will not be tampered with for a week — until next week's lab. You should see the bacteria reproduce and spread out over the agar in each petri dish. They will appear as a beige or white filmy covering over the agar. It will take at least a week (maybe a little longer) for the *E. coli* to cover most of the agar.

7. Place the cotton swab into a jar (or other small container) of bleach. Let it soak for about an hour in the bleach to kill the bacteria. After that the cotton swab can be placed in the garbage. You may have a couple more sterile swabs in case you need to repeat the procedure.

8. Now, for fun, take a sample from your mouth (or other place like the kitchen or bathroom counter) with another sterile swab and smear it onto a fresh petri dish agar. Set it aside for a week to see what grows.

9. Keep the tube with the original *E. coli* bacteria in case you need to go back and repeat the procedure for whatever reason. Be sure to keep the cap on the tube so as not to contaminate it.

## Laboratory Report (20 points possible)

You will receive full credit for this laboratory assignment when you complete the procedure. You will complete a report on this assignment next week.

## Matching

A. Fetus          B. Conception          C. Embryo          D. Diploid

1. _____ The haploid egg and haploid sperm cells unite to form one _____ cell, the zygote.

2. _____ The zygote divides, forming a many-celled _____.

3. _____ After eight weeks of development, a human is called a _____.

4. _____ We are fully human at _____ as a zygote, as King David says in Psalm 51:5.

## Fill in the Blank

5. We have _____ (also called _____) genes in common with many animals.

6. These genes start the development of arms, legs, eyes, and other important _____ _____.

7. Bacteria do not have to divide a nucleus because they do not have one. They divide by _____ _____.

8. The stage in between cell divisions is _____.

9. In this stage, _____ and _____ replicate, and the cell grows and carries out specialized functions.

10. _____ is the distribution of the chromosomes into 2 daughter nuclei.

11. The division of the cell itself into two daughter cells is called _____.

12. During the first stage of mitosis, called _____, the membranes around the _____ come apart and _____ _____ attach to the chromosomes.

## Short Answer

13. In prophase:

## Matching

A. Telophase      B. Metaphase      C. Meiosis      D. Anaphase

1. _____ The chromosomes are pulled toward the middle of the cell during this.

2. _____ As the chromosomes keep being pulled away from the middle of the cell toward the ends of the cell, the stage is called what?

3. _____ The chromosomes are at the opposite ends of the cell and the nuclei reform during this.

4. _____ The process of forming haploid cells is called what?

## Fill in the Blank

5. In this process, a cell goes from 2n to _____. Then this cell divides to form ____ cells that are each _____ n.

6. Human diploid cells have a total of _____ chromosomes.

7. In prophase I of meiosis, the homologous chromosomes go through a process called _____ where they come together to form a _____.

8. During meiosis, the egg _____ 3 out of 4 copies of each chromosome, leaving only _____ egg.

9. Sometimes sister chromatids undergo _____ _____ whereby they exchange genes with each other. This results in greater variation and often enhances the _____ ability of the offspring.

10. The egg cell retains most of its _____, which becomes _____ for the zygote during its early development.

11. _____ are circular bacteria DNA that are not part of the main chromosome.

12. When a bacterium cell takes up this DNA from another bacterium cell and makes it part of its own it is called _____.

13. _____ is when a virus inserts DNA into a bacterium cell.

14. An example is when a virus carried the DNA that coded for human _____ into an E. coli cell.

15. _____ _____ is where DNA is taken from one organism and intentionally inserted into that of another organism.

16. DNA formed as a combination of DNA from different species is called _____ DNA.

## Short Answer

17. In Jeremiah 1:5 the Lord told Jeremiah:

## Laboratory 18: Diverse Products of Protein Translation at Different Stages of Development

**Lab Notes:**

### REQUIRED MATERIALS

- ☐ Microscope (from the supply kit)
- ☐ Prepared slide of Fern Life Cycle (from the supply kit)
- ☐ Prepared slide of Moss Life Cycle (from the supply kit)
- ☐ Petri dishes with *E. coli* cultures from Laboratory 17 — **Optional**
- ☐ Petri dishes with other samples that you may have chosen to do — **Optional**
- ☐ Antibiotic discs (ampicillin, erythromycin, neomycin, and penicillin) (from the supply kit) — **Optional**
- ☐ Tweezers — **Optional**
- ☐ 100 ml beaker — **Optional**

### INTRODUCTION

Ferns and mosses are dramatically different at different stages of their life cycle. Most know ferns as the narrow stem with many narrow branches (called fronds) coming out to the sides. These are used in flower bouquets.

Under the fronds, round dark spheres called sori (plural for sorus) develop. They undergo meiosis to produce spores that are haploid (have 1 of each chromosome). The spores are released and grow into a leaf-like structure (called a gametophyte) that is found on the ground. Part of the gametophyte produces haploid egg cells and part of it produces haploid sperm cells. When the nucleus of an egg cell unites with the nucleus of a sperm cell, a zygote develops through many cycles of mitosis into a new fern called a sporophyte. These structures in the life cycle of a fern are very different from each other but they are still the same organism with the same DNA. It is just that different parts of the DNA produce different mRNA at different stages of the lifecycle.

The life cycle of a moss is similar but different from that of a fern. The leafy stalks of moss are the sporophytes (diploid). On the top of the stalks, structures develop called sporangia. Within these structures haploid male and female spores develop by meiosis. When released, the spores develop into separate male and female gametophytes which produce haploid egg cells and sperm cells. They unite their nuclei to form zygotes that form new diploid sporophytes that look like the long stalks of moss.

**Optional Enrichment Exercise**

This section refers to the continuation of the optional enrichment exercise from laboratory 17. The *E. coli* bacteria form a white- or beige-colored film as it grows over the agar in the petri dishes. If an antibiotic kills or keeps the *E. coli* from dividing, a clear spot is formed around the antibiotic disc in the film covering the agar. In this exercise, you are placing paper discs soaked in the antibiotics ampicillin, erythromycin, neomycin, and penicillin onto the *E. coli* in each of the four sections that you marked last week on the cover of the petri dish. After a few days, you should see a clear region around the paper disc if that antibiotic is killing the bacteria cells. Neomycin reacts by disrupting the ribosomes from producing proteins. Ampicillin penetrates the outer membrane of the *E. coli* cells and inhibits the enzyme transpeptidase that is needed for cell wall production. Some strains of *E. coli* produce an enzyme called erythroesterase that inactivates erythromycin. How would you know if the erythromycin was inactivated?

## PURPOSE

DNA transcription and translation can be very different at differences stages of life of an organism. Think about how the proteins that result in a tadpole must be different from the proteins of a mature frog even though they are the same organism. Think of how different you were at one year old as contrasted to now. In this exercise, the stages in the life cycles of ferns and mosses are observed.

The optional enrichment exercise concludes the work from last week. The goal is to observe whether or not each of the antibiotics disrupted the process of DNA translation or the products of DNA translation. As well, a goal is to determine what grew from the optional samples (such as your mouth) collected last week.

## PROCEDURE

1.  Examine the fern life cycle prepared slide under low power. These structures are large enough to be observed with the low power. Identify the sporophyte (large stalk) stage and the gametophyte (looks like a flat leaf) stages. Point these out to your teacher and describe the fern life cycle while referring to each structure. Look up and describe the uses for ferns.

2.  Examine the moss life cycle prepared slide under low power. These structures are also large enough to be observed with the low power. Identify the sporophyte (a stalk where the leaf-like structures do not go all the way to the top) stage and the sporangia at the top of the stalks.

Identify the gametophyte (look like a stalk with leaf-like structures all the way to the top) stages. Point these out to your teacher and describe the moss life cycle while referring to each structure. Look up and describe the uses for moss.

**Optional Enrichment Exercise**

1. Examine and describe each of the petri dishes from last week. Was the *E. coli* successful in covering the agar in the 2 petri dishes?

2. For the 2 petri dishes with *E. coli*, carefully place an antibiotic paper disc in each of the 4 sections of the agar. You marked the 4 sections last week on the cover of the petri dish. Place the antibiotic discs so that the markings on them are facing up so that you can read them. Be sure not to touch the discs with your hands or to lay them down anywhere except on the agar.

3. Place some antibiotic discs on the optional growths if you made them.

4. Place the petri dishes aside for a few days in a warm place.

5. After a few days, examine and describe what happened around each antibiotic disc. Make a drawing or take a picture of your petri dishes and submit it as part of your report.

6. Make a chart like the following showing the results of each antibiotic, and a possible explanation for each result. Go back to the Introduction and you will see helps for ampicillin, erythromycin, and neomycin. You may need to look up the effects of penicillin.

| Antibiotic | Effect |
|---|---|
| Ampicillin | |
| Erythromycin | |
| Neomycin | |
| Penicillin | |

7. What happened to your optional sample? Write out a possible explanation for your results (remember this one is optional).

**Laboratory Report** (20 points possible)

Step 1

Step 5

Step 6

Step 7

## Fill in the Blank

1.  The *R* allele produces a protein that causes the peas to be _____.

2.  A pea plant that has the alleles *rr* produces _____ peas.

3.  *RR* plants have _____ peas.

4.  *Rr* plants have _____ peas.

5.  The genetic pattern *RR*, *Rr*, or *rr* is called the _____.

6.  The physical appearance of the peas is the _____.

## Fill in the Diagram

7.  Punnett square of *Rr* x *Rr*

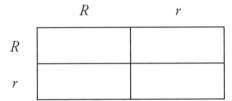

## Matching

A.  Long          B.  Short          C.  Homozygous          D.  Heterozygous

8.  _____  When an *LL* tom cat breeds with an *ll* female cat, the kittens will have the genotype *Ll* and have this kind of hair.

9.  _____  The condition whereby both alleles are the same.

10.  _____  The condition whereby the alleles are different.

11.  _____  When an *Ll* tom cat breeds with an *ll* female cat, 50% of the kittens will have short hair and 50% of the kittens will have this kind of hair.

## Matching

A. Mildly frizzled        B. Very frizzled        C. Normal

1. _____ $F^+F^+$ chickens have this kind of feathers.

2. _____ $F^+F^f$ chickens have this kind of feathers.

3. _____ $F^fF^f$ chickens have this kind of feathers.

## Fill in the Blank

4. The + symbol indicates the most _____, or _____ type, trait.

5. The use of $F$ for both alleles indicates that neither is _____ over the other.

6. An $F^+F^+$ chicken crossed with an $F^fF^f$ chicken will produce chickens with the genotype _____ and the phenotype _____ _____.

7. When an $F^+F^f$ chicken is crossed with an $F^+F^f$ chicken, the resulting chicks will be _____ % normal, _____ % mildly frizzled, and _____ % very frizzled.

## Short Answer

8. What did Gregor Mendel notice about plants?

9. What did this help him predict?

## Laboratory 19: Human Genetics

**Lab Notes:**

### REQUIRED MATERIALS

☐ Pencil and paper
☐ PTC taste test paper (from supply kit)

### INTRODUCTION

Many human traits are dominant or recessive. Their inheritance pattern can be traced with a pedigree chart. Sometimes when a very expensive dog is purchased, the seller is expected to provide the dog's pedigree. This consists of the identification of the breed of the dog's parents and the genetic make-up of important traits — such as patterns of fur color, temperament, and other traits, and are the dog's alleles dominant or recessive for particular genes.

Some people have free-hanging earlobes and some have attached earlobes. Free-hanging earlobes are a dominant trait and attached earlobes are recessive. When a trait is expressed in about ¾ of a population, it is considered to be dominant. The genotype is the alleles of the gene and the phenotype is the physical appearance. The allele for free-hanging earlobes (E) produces mRNA that codes for a protein that causes the earlobes to be unattached. The other form of the gene (allele, e) produces a protein that causes the earlobes to be attached. If a person has the genotype EE or Ee, they will produce the protein that makes unattached earlobes. A person who is ee will produce the form of the protein that results in attached earlobes.

The ability to roll the tongue is a dominant trait and the lack of ability to roll the tongue is a recessive trait.

The hair line called a widow's peak is a dominant trait and the straight hair line is a recessive trait.

The presence of freckles is a dominant trait and the lack of freckles is a recessive trait. It is a case where most people do not have freckles. This happens when a group becomes isolated from the majority of people who have freckles. This is evident, as well, in populations that have predominantly blue eyes (also a recessive trait). Most people that live far north of the equator tend to have blue eyes. Those that live closer to the equator tend to have brown eyes (a dominant trait). This is because of how they are distributed rather than caused by cold or warm climates. People with blue eyes migrated farther north after the Ice Age. Freckles and brown eyes are caused by proteins produced by the dominant genes. The lack of freckles and blue eyes are due to these genes being modified so that they produce modified proteins.

Dimples are a dominant trait and the lack of dimples are a recessive trait. So, if you have dimples you got them from your parents.

If you are a PTC taster, you have a dominant gene. In this exercise, you will taste small strips of paper soaked in phenylthiocarbamide. It has a bitter taste. A similar compound gives broccoli, cabbage, and Brussels sprouts their distinct taste. Usually some like these vegetables and others do not like them. Perhaps the tasters do not like the bitter taste and non-tasters cannot tell the difference.

Right-handedness is a dominant trait and left-handedness is a recessive trait.

## PURPOSE

This laboratory exercise provides practice in tracing back lineages of dominant and recessive traits and working with human pedigree charts.

## PROCEDURE

1. Construct a pedigree chart like the following.

Generation 1: Heterozygous (1 dominant allele and 1 recessive allele) male marries homozygous (2 dominant alleles) female

Generation 2: Offspring #1 (homozygous female); #3 (heterozygous female); and #4 (heterozygous male). #3 female marries #2 male and #4 male marries #5 homozygous female

Generation 3: #2 and #3 (from Generation 2) have 4 children (#1 homozygous recessive male; #2 homozygous dominant female; #3 homozygous recessive female; and #4 heterozygous female).

A. The circles stand for females and squares stand for males. In the samples chart, the top circle is the mother and the top square is the father. The second row stands for the children. This is you and your siblings. If you are a male and you have 2 sisters, the second row will have 1 square and 2 circles.

B. For each individual with a dominant trait, the circle is not colored in. For each individual with a recessive trait, the circle or square is colored in.

2. Make a chart. For example, if the offspring are 1 male and 2 females, the chart will look like this.

father and mother
have brown eyes

brown
eye
male

blue
eye
female

brown
eye
female

Example: = male with 2 dominant alleles for brown eyes (dominant)

= male with 1 allele for brown eyes and 1 allele for blue eyes (recessive)

= male with 2 alleles for recessive blue eyes

3. If both parents have free-hanging earlobes, the son has free-hanging earlobes, one daughter has free-hanging earlobes, and the other daughter has attached earlobes, the pedigree chart looks like this.

free hanging ear lobes dominant allele
attached ear lobes recessive allele

both father and mother
have free hanging
ear lobes

free
hanging
ear lobes

free
hanging
ear lobes

attached
ear lobes
(has to have both
recessive alleles)

The parents have to have a dominant and a recessive allele for the second daughter to have 2 recessive alleles. The son and the first daughter can be either with 2 dominant alleles or with 1 dominant allele and 1 recessive allele.

4. The second daughter will be ee (2 recessives) and the others with free-hanging earlobes will be EE or Ee. The parents have to be Ee because that is the only way that the second daughter could be ee (she got 1 e from each parent). The son and first daughter could be EE or Ee. You would have to wait until they got married and see what kind of earlobes their children have.

5. Construct a separate pedigree chart for your family for each of the following: tongue rolling, widow's peak hairline, freckles, blue or brown eyes, dimples, and PTC tasters. When you try to taste the PTC paper, you may need to wad up 2 or 3 pieces and chew them a bit. Do not swallow them. They are just not on your diet. Color in or do not color in the squares and circles depending upon whether or not each person has the dominant trait or the recessive trait. This gives 6 charts in addition to the one for the earlobes. The chart only shows the phenotype (the physical appearance). For each chart see how many genotypes you can come up with (EE, Ee or ee for earlobes). Just for fun, check to see if the PTC non-tasters like broccoli and Brussels sprouts. Do they taste bitter to the PTC tasters?

6. Your lab report consists of the 7 pedigree charts and the genotypes that you can determine.

**Laboratory Report** (20 points possible)

Step 2 Pedigree Chart

Step 3 Pedigree Chart

Step 4 Pedigree Chart

Family Pedigree Chart for Tongue Rolling

Family Genotypes for Tongue Rolling

Family Pedigree Chart for Widow's Peak Hairline

Family Genotypes for Widow's Peak Hairline

Family Pedigree Chart for Freckles

Family Genotypes for Freckles

Family Pedigree Chart for Blue or Brown Eyes

Family Genotypes for Blue or Brown Eyes

Family Pedigree Chart for Dimples

Family Genotypes for Dimples


Family Pedigree Chart for PTC Tasters


Family Genotypes for PTC Tasters

## Short Answer

1. The Law of Independent Assortment states that:

## Matching

> A.  Green (*Ry*)    B.  Crossed    C.  Yellow (*RY*)

2. _____ An *RRYY* pea plant will have all round peas of this color.

3. _____ An *RRyy* pea plant will have all round peas of this color.

4. _____ *RrYy* pea plants _____ with each other to get $F_2$ plants can have *RY, Ry, rY,* and *ry* pollen grains and eggs.

## Fill in the Diagram

5. Draw the Punnett Square for a cross between *RrYy* x *RrYy*.

Punnett Square *RrYy* x *RrYy*

| | | | |
| --- | --- | --- | --- |
| | | | |
| | | | |
| | | | |

## Fill in the Blank

6. The observed results may not exactly match the expected results because of _____ events.

7. If the probability of getting different results from the observed and expected is less than _____ percent, the results are not _____ _____ from each other.

## Fill in the Blank

1. The genes for body color and eye surface for fruit flies are _____, meaning that they are on the same _____.

2. You would expect a _____ ratio of wild type to ebony rough from a ++er x ++er cross.

3. However, the resulting percentages are about _____ wild type (rather than 75%), _____ ebony rough (instead of 25%), _____ ebony with normal eyes, and _____ rough eyes and normal gray bodies.

4. The results do not come out exactly as expected because of the phenomenon called _____ _____.

5. When the homologous chromosomes come together, they are said to _____.

6. This forms a _____, which means four.

7. Under the microscope, it still looks like two _____, but there are really four.

8. A picture of an organism's chromosomes is a _____.

9. The 23rd pair of human chromosomes looks like an _____ and an _____ in females.

10. The 23rd pair of human chromosomes looks like an _____ and a _____ in males.

11. Some _____ on the section of the _____ chromosome are not on the _____ chromosome.

12. This is why _____ and _____ _____ are far more common in males than females.

## Matching

A. Mistakes    B. Amniocentesis    C. Missing    D. Abnormalities

13. _____ The process of drawing out amniotic fluid around a fetus is called:

14. _____ Fetal cells from the fluid are used to make a karyotype to detect:

15. _____ The evolution model says these abnormalities are evolutionary:

16. _____ The easiest defect to see is this kind of chromosome or too many chromosomes.

## Short Answer

17. Many individuals that were declared defective by karyotypes are leading genetically normal healthy lives. This goes back to the basic biblical concept that we are far more than a bag of chemical reactions like a test tube in a laboratory. Explain the connection with God's creation:

## Laboratory 20: Dihybrid Test Cross with Corn

**Lab Notes:**

### REQUIRED MATERIALS

- ☐ Note pad and pencil
- ☐ Image of corn cob with kernels (seeds) produced from a cross of R/r Su/su x r/r su/su

### INTRODUCTION

The kernels on a corn cob are the seeds. They are produced when the pollen (produced by the tassel of the corn) unites with the egg in the flower. R stands for the dominant allele that produces purple kernels (seeds) and r stands for the recessive allele that produces yellow kernels. Would you eat purple corn?

Su stands for the dominant allele for another gene that produces a lot of starch in the kernel, which gives the kernel a firm shape; su stands for the allele of the same gene that causes the kernel to be very sweet with much less starch. These kernels appear wrinkled.

R/R Su/Su kernels are purple and firm. R/r Su/su kernels are also purple and firm because R and Su are dominant.

If you have a corn cob with purple firm (starchy) kernels, how do you know if their genotype is R/R Su/Su or R/r Su/su? To determine this, you need to do a test cross. This is done by planting several kernels from your cob and later collecting the pollen from the tassels that develop from the new plants. Sprinkle this pollen onto the flowers from plants that grew from r/r su/su kernels. When the new cobs grow on the resulting plants, identify and count the kernels on one of the cobs. If all of the kernels on the resulting cob are all purple and firm, the original corn was R/R Su/Su because when it was crossed with r/r su/su all of the resulting kernels would be R/r Su/su, which would have purple firm kernels. If the original corn was R/r Su/su, the resulting kernels on the new cob would be as shown in this Punnett Square.

R/r Su/su x r/r su/su

| | r su | r su | r su | r su | |
|---|---|---|---|---|---|
| R Su | Rr Susu | Rr Susu | Rr Susu | Rr Susu | purple firm |
| R su | Rr susu | Rr susu | Rr susu | Rr susu | purple wrinkled |
| r Su | rr Susu | rr Susu | rr Susu | rr Susu | yellow firm |
| r su | rr susu | rr susu | rr susu | rr susu | yellow wrinkled |

This is a 4:4:4:4 ratio

If you cross a R/r Su/su x R/r Su/su you would get a 9:3:3:1 ratio as in this Punnett Square

### R/r Su/su x R/r Su/su

|       | R Su    | R su    | r Su    | r su    |
|-------|---------|---------|---------|---------|
| R Su  | RR SuSu | RR Susu | Rr SuSu | Rr Susu |
| R su  | RR Susu | RR susu | Rr Susu | Rr susu |
| r Su  | Rr SuSu | Rr Susu | rr SuSu | rr Susu |
| r su  | Rr Susu | Rr susu | rr Susu | rr susu |

9 have an R and a Su = purple, firm
3 have rr and a Su = yellow, firm
3 have a R and susu = purple, wrinkled
1 has rr and susu = yellow and wrinkled

You would get a ratio of 9:3:3:1, meaning that 9 out of 16 would be purple and firm (R/R Su/Su or R/R Su/su or R/r Su/Su or R/r Su/su); 3 out of 16 would be purple and wrinkled (sweet) (R/R su/su or R/r su/su); 3 out of 16 would be yellow and firm (r/r Su/Su or r/r Su/su); and 1 out of 16 would be yellow and wrinkled (r/r su/su).

## PURPOSE

This exercise will provide direct experience in studying the results of a dihybrid test cross. Sometimes looking at the results of an experimental procedure is a more effective learning tool than a written explanation. The explanation of the results of the test cross in the introduction may have left you scratching your head (or someone else's head).

## PROCEDURE

1.  Carefully count and write down how many of each kind of kernel you find on the cob in the above photos. Be clever in your technique so that you count every kernel only once and none are missed. From this information, fill in this chart.

### Results of your counts of the corn cob

| corn kernels     | count | divide count by yellow wrinkled |
|------------------|-------|---------------------------------|
| purple firm      |       |                                 |
| purple wrinkled  |       |                                 |
| yellow firm      |       |                                 |
| yellow wrinkled  |       |                                 |

The ratio of purple firm:purple wrinkled:yellow firm:yellow wrinkled is

_____ : _____ : _____ : _____

| purple firm | purple wrinkled | yellow firm | yellow wrinkled |
|---|---|---|---|

2. Were all of the kernels on the cob purple and firm with no purple wrinkled, yellow firm, or yellow wrinkled? If this is the case, the original corn cob would have been R/R Su/Su. Describe in your own words in a complete sentence why this would be the case.

3. To find your ratios, first count the number of kernels that are yellow wrinkled. These should be the last number (1) in the 9:3:3:1 ratio. Count the purple firm kernels and divide this number by the number of yellow wrinkled kernels. This should give the 9 in the ratio. Then count the purple wrinkled kernels and the yellow firm kernels and divide each of these numbers by the number of yellow wrinkled kernels. These should give the two number 3's in the ratio. Your numbers will not come out to be exactly 9:3:3:1. This is normal for any scientific study. If you take courses in statistics in the future, they will cover methods of determining which are reasonable differences and which are not.

4. Did you get close to a 9:3:3:1 ratio as described in the introduction to this lab? If so, the original cob would have had all purple firm kernels with the genotype R/r Su/su. Describe in your own words in a complete sentence why this would be the case.

5. How did you make sure that all of the kernels were counted and that each was counted only once?

6. If you think that your results are close enough to a 9:3:3:1 ratio, why would it not be exactly a 9:3:3:1 ratio?

7. Does this make more sense to count the kernels than just read the introduction?

**Laboratory Report** (20 points possible)

Number of purple firm kernels _____

Number of purple wrinkled kernels _____

Number of yellow firm kernels _____

Number of yellow wrinkled kernels _____

Step 3

Step 4 — To find the ratios of kernel types, divide each number of kernels by the number of yellow wrinkled kernels. It will look like this pf/yw: pw/yw: yf/yw: yw/yw. This gives yw/yw to be 1. You are comparing it to the 9:3:3:1 ratio.

Step 5

Step 6

Step 7

## Short Answer

1. Aristotle used the terms genus and species to mean:

## Fill in the Blank

2. Genesis states that organisms were created in groups called _____.

3. The phrase "fixity of the species" means that the _____ is the same as the _____ of an organism.

4. In 1686, _____ _____ was the first one to use the word _____ as a biological term.

5. In the 1700s, _____ _____ placed similar organisms in the same _____.

6. Organisms that interbreed and have offspring are in the same _____.

7. There is _____ genetic variation today than right after the Flood of Noah.

## Matching

A. Translocations     B. Point     C. Transposons     D. Codon

8. _____ These mutations change the base pairs of DNA.

9. _____ Because of redundancy in the genetic code, more than one of this combination codes for the same amino acid.

10. _____ This is where parts of chromosomes are moved around to different locations in the chromosome or between chromosomes.

11. _____ These are sections of chromosomes that are moved around.

## Matching

A. Gene    B. Duplication    C. Polyploidy    D. Deletions

1. _____ This occurs where there is more than one copy of a gene.

2. _____ If a gene were to be changed enough, a new one of these could result.

3. _____ These are always harmful because they result in a missing protein in a cell.

4. _____ This is where there are more than two of each chromosome.

## Fill in the Blank

5. This occurs more often in _____ than in _____.

6. If a plant species with four pairs of chromosomes is crossed with a similar plant species with three pairs of chromosomes and they produce a plant with seven pairs of chromosomes, it is called an _____.

7. Polyploidy plants usually have larger _____ and _____ and are sometimes _____ (like popular watermelons).

8. Second Samuel describes the Philistine soldier Goliath as having six fingers on each hand and six toes on each foot. Today we know this trait to be a mutation called _____.

9. _____ _____ is where the environment causes organisms with certain genetic alleles to survive better than others.

10. _____ _____ is where humans deliberately select certain combinations to interbreed to produce offspring with particular alleles.

## Short Answer

11. Who was Dr. Walter Lammerts?

## Laboratory 21: Genetic Variation

**Lab Notes:**

### REQUIRED MATERIALS

☐ Paper and pencil
☐ Access to insects or pictures of insects

### INTRODUCTION

This chapter is about a number of ways that modifications of DNA can occur that bring about variations in the genotypes of the next generation. There are limits as to the nature and amount of variations of DNA that can take place. If the DNA is changed such that critical proteins are produced that cannot function, the organism dies.

You have certainly noticed that you have some differences from your parents and siblings. What if your parents told you that you were going to have a baby brother who would be hairy like a monkey and have webbed feet? That is silly because it does not happen.

Can a German shepherd give birth to a collie? This does not happen. But a German shepherd can mate with a collie and have puppies that look like part German shepherd and part collie. The German shepherd does not have some of the alleles that are needed to produce a collie — but a collie does. It appears that the dogs that were released from Noah's Ark became all of the other dogs. That was thought to be impossible at one time, but after the mechanisms introduced in this chapter and some others were better understood, it seemed reasonable. This is not the same as evolution because it is making subtle changes in DNA that was already there. In spite of these variations, they are still all dogs.

In this lab exercise, you are going to look at a number of insects and see some similarities and some differences. In the Book of Genesis, God commanded the life forms to reproduce after their kinds. This created distinct groups that could not cross over between each other to produce offspring. But then God allowed for very limited variations within each kind so that as they spread out over the earth they could survive in very different environments.

Biological classification is based upon a system of dividing organisms down into smaller and smaller groups based upon their similarities and differences. This is the system for insects.

## DOMAIN EUKARYA
### (have true nuclear membranes)
## KINGDOM ANAMALIA
## PHYLUM ARTHROPODA
## CLASS INSECTA

The class Insecta is broken down into orders which are studied in this exercise. To complete the classification levels, orders are broken down into families, which are broken down into genera, which are broken down into species.

## PURPOSE

This exercise will give you experience looking at insects that appear to be in very distinct groups and insects that are very similar to each other that could possibly be in the same kind of genus.

## PROCEDURE

1.  What you do here will depend upon your access to insect specimens. If you are doing this exercise in a region where it is spring with many insects hatching out, you should go out and see how many you can find that are listed below. If you have snow on the ground and spring seems like it will never come, then you will need to look up pictures of the insect. You will probably have to look up some, no matter where you are. If available, you should be able to find some in a pet shop.

2.  Find pictures (the Internet should be very helpful) or specimens of insects in the following orders.

    A.  Lepidoptera — butterflies and moths
    B.  Diptera — flies
    C.  Ephemeroptera — mayflies
    D.  Hymenoptera (winged) — winged ants, wasps, and bees
    E.  Odonata — dragonflies and damselflies
    F.  Isoptera — termites (I hope you do not find these specimens)
    G.  Hymenoptera — ants
    H.  Orthoptera — cockroaches, walking sticks, grasshoppers, crickets
    I.  Coleoptera — beetles
    J.  Hemiptera — true bugs

3.  For each of these orders, try to find several examples that you can look at. For each order, try to list those together that you think could have been in the same kind of creation. This means that you would expect that they are

similar enough to each other where they could interbreed and produce offspring that could again produce offspring. Each order is going to have insects that would not be in the same kind. Some think that in some cases, perhaps those in the same family or genus may be in the same kind. You could only tell for sure if you did breeding experiments with them. The purpose of this exercise is to gain experience dealing with this concept rather than to actually identify which are in the same kind. This is difficult even for well-trained and experienced entomologists (insect specialists).

4. Prepare your report in an orderly fashion so that someone else looking at it will understand it. It is your option whether you include pictures in your report or not. That will depend on what resources you have available.

## Laboratory Report (20 points possible)

On the chart below, list insects that you found or found pictures for in their proper order. Try to list at least 3 for each order. This demonstrates the great and fascinating variations that God has produced in His creation.

Lepidoptera

Diptera

Ephemeroptera

Hymenoptera

Odonata

Isoptera

Orthoptera

Coleoptera

Hemiptera

## Short Answer

1. What did Gregor Mendel realize?

## Fill in the Blank

2. Genomics is the study of the sequences of _____ that make up _____ and entire _____.

3. Bioinformatics involves the use of _____ _____ to analyze and interpret the sequence of bases.

4. Since 2003, the techniques for _____ _____ have become more efficient and economical.

5. By sequencing a person's DNA, it is possible to design _____ specific to that person's particular needs.

6. An SNP is an area in DNA where there is a different _____ than in the majority of people.

7. SNP stands for _____ _____ _____.

8. The base that differs is called an _____.

9. The genetic makeup of a person is the person's _____.

## Matching

　　　　　A. Kind　　　　　B. Haplotype　　　　　C. Genomics　　　　　D. HapMap

10. _____ A sequence of SNPs is called this.

11. _____ A section of DNA containing SNPs is called this.

12. _____ The study of similarities and differences of genomes is called comparative what?

13. _____ Species that on rare occasions produce hybrids are usually considered to be in the same _____ of creation.

## Matching

A. Amniocentesis      B. Killed      C. Clone      D. Functions

1. \_\_\_\_\_ It was found that about 2/3 of the genes involved in human cancer are also found in fruit flies, which indicates common what rather than common descent or ancestry?

2. \_\_\_\_\_ Genomics is a much better technique than a test called this.

3. \_\_\_\_\_ Even if a fetus has severe genetic weaknesses, it should not be _____ but cared for with God's grace. Many such individuals have been used mightily by God in the lives of others.

4. \_\_\_\_\_ An organism that is genetically identical to another.

## Fill in the Blank

5. Cloning usually involves taking _____ from a _____ egg cell and placing it into an egg that had its _____ removed.

6. The first mammal to be cloned was a _____ named _____.

7. An example of a plant that reproduces asexually (without egg and sperm cells), producing clones, is the _____ _____.

8. An invertebrate animal that naturally forms clones when a chunk of its body falls off is the _____ _____.

## Short Answer

9. What is your personal view concerning cloning people?

## Laboratory 22: Plants

### REQUIRED MATERIALS

- ☐ Microscope (from the supply kit)
- ☐ Prepared slide of *Ranunculus* root cross section (from the supply kit)
- ☐ Prepared slide of *Ranunculus* stem cross section (from the supply kit)
- ☐ Prepared slide of *Ficus* leaf cross section (from the supply kit)

### INTRODUCTION

This chapter deals with genomics (the study of and application of DNA sequencing). This lab exercise is jumping ahead to the next chapter that deals with taxonomy in general and plant taxonomy in particular. These are based upon the physical characteristics of living organisms and increasingly upon the similarities in their genomes.

When living forms are referred to in Genesis, it usually refers to animals. Plants are considered to be quite different. They were the sole source of food until after Noah and his family got off the Ark, and are also referred to as food in the New Jerusalem.

Plant structures, which are the products of their genomes, are very different than those of animals. As well, seaweed are not classified as plants because their structures are very different.

All of the basic life functions are carried out by plants. They have mitochondria for respiration and chloroplasts (that animals do not have) for photosynthesis. Plants do not have brains and complex nervous systems even though there is some communication between plant cells. Plants are not aware of being alive like animals.

The genus *Ranunculus* includes flowers called buttercups. It used to be popular to say that if a buttercup flower were held under someone's nose and the person sneezed, that person is in love with the one holding the flower. Do not use this as a means to decide who to propose marriage to. In fact, the *Ranunculus* are slightly poisonous. Roots are important to bring water and minerals up into the plant from the ground and the stem brings water and minerals up into the plant. Another plant studied in this exercise is the leaf of the *Ficus* tree. These plants are dicots, meaning that their seed has two leaf structures and their leaves are broad with veins that branch out rather than being parallel.

**Lab Notes:**

Taxonomy is the art of classifying living things. This is a means of organizing so that those that are similar can be grouped together and those that are different can be placed in separate groups. This is not as easy as it might sound, as you will see.

## PURPOSE

1. In this exercise, you will see the phenotypes (physical characteristics) of roots, stems, and leaves produced by the genotypes of some common plants. This will be helpful in understanding their taxonomy in the next chapter.

2. You will gain additional experience in using the microscope.

## PROCEDURE

1. Place the prepared microscope slide of *Ranunculus* root cross section on the stage of the microscope and examine it with low power. A cross section is a very thin section cut across, separating the root into a top and bottom portion. At least 2 cuts are made so that you end up with a thin round section on the microscope slide. The nice thing is that someone else cut, mounted, and stained it for you. There are 3 main regions of the root. The outer part is the **epidermis** (outer covering). The middle of the root is composed of a number of circles which are tubes cut across. The larger circles are cut **xylem** tubes that carry water and minerals. The smaller circles are cut **phloem** tubes that carry sugar water produced in the upper green part of the plant down into the roots. The portion in between the middle and the outer portion is the **cortex**, which gives physical support to the root. Point out the epidermis, cortex, xylem tubes, and phloem tubes and their functions to your teacher.

2. Place the prepared microscope slide of the *Ranunculus* stem cross section on the stage of the microscope and examine it with low power. The stem is the above-ground support for the flower. It has to hold up the flower and carry water and minerals from the roots to the rest of the plant. Stems are very different from roots because they have to be strong enough to hold up the plants. The thin outer covering is the **epidermis** like in the root. The phloem and xylem tubes appear as columns that appear as circles in the cross section. They are just under the epidermis. Together, the phloem and xylem are called **vascular tissue** (or veins). The central portion of the stem is composed of many larger circles called **pith**. The pith gives additional support to the stem. It also gives support to the stem when it is bent by blowing wind. Point out the epidermis, vascular tissue, and pith and their functions to your teacher.

3. Place the prepared microscope slide of *Ficus* leaf cross section on the stage of the microscope. *Ficus* is an ornamental tree found in many landscapes.

4. The leaf has a thick portion down the middle where the **veins** (phloem and xylem) carry water, minerals, and sugar water into the leaf. As you look at the slide, the leaf looks like a "V" shape with the thick vein at the bottom of the V and the sides of the leaf going up on both sides. The top and bottom of the leaf has a thin layer called the epidermis. Above the **lower epidermis** are the **veins** that go out into the leaf. Under the **upper epidermis** are **mesophyll cells** where photosynthesis takes place. Point out the central vein, upper epidermis, mesophyll cells, lower epidermis, and the veins above the lower epidermis and their functions to your teacher.

5. Take a slice of old moist bread and an orange cut in half and place them somewhere where they can get moldy without being eaten by a pet, insects, or your little brother. You will use these to study mold in next week's lab.

**Laboratory Report** (20 points possible)

Point out the following on the microscope slides and the assigned structures to your teacher.

Male pine cone

Female pine cone

Pine root cross section

Pine stem cross section

Your grade is the percent of correct identifications out of a possible 20 points.

## Matching

A. Descent          B. Traits          C. Eukarya          D. Classification

1. _____ Taxonomy is the study of:

2. _____ Some biologists say they classify organisms according to evolutionary:

3. _____ Traditionally, plants are classified according to their physical:

4. _____ Plants are in the domain:

## Fill in the Diagram

5.

## Fill in the Blank

6. Domains are divided into _____. Plants are in the kingdom _____.

7. The categories within a kingdom (in order from larger to smaller) are _____, _____, _____, _____, _____, and _____.

8. The first letter in a genus name is always _____. The first letter in a species name is always _____. Both names are always _____ or written in _____.

9. In the name *Homo sapiens*, *Homo* is the _____ name and *sapiens* is the _____ name.

10. Plants are divided into the broad categories of _____ and _____ plants.

11. Vascular plants have _____ and other vessels that carry nutrients and water throughout the plants.

12. Vascular plants are grouped into those that produce _____ and those that do not.

13. Non-vascular plants are called _____.

14. They have a structure called a _____ because by cell division, it produces _____ (egg and sperm cells).

## Matching

A. Lycophyta    B. Pterophyta    C. Mycorrhizae    D. Psilophyta

1. _____ Seedless vascular plants include the phyla psilophyta, lycophyta, sphenophyta, and:

2. _____ Whisk ferns are in this phylum.

3. _____ Whisk ferns have a mutualistic relationship with this kind of fungi where the fungi help move water and nutrients throughout the plants.

4. _____ Club mosses are in this phylum.

## Fill in the Blank

5. Club mosses are not really _____. The forms living today are not very tall, but there are fossils that show huge forests of these that stood up to _____ feet tall.

6. Horsetails are in the phylum _____.

7. The stems of horsetails are _____ and their outer coverings contain _____, which is also a major part of glass.

8. Ferns are in the phylum _____. They have clusters of narrow leaves and require a lot of moisture.

9. Vascular plants that produce seeds are in the phyla _____ and _____/ _____.

10. Conifers produce seeds in _____ and have _____- _____ leaves.

11. Conifers are _____, meaning that they do not lose their leaves in the fall except for _____, whose leaves turn yellow and drop to the ground.

12. The phylum Anthophyta/Magnoliophyta is divided into the class _____ and the class _____.

13. Monocots include _____, _____, _____, and _____. The veins in their leaves are _____, and their flower parts come in _____ or multiples of 3.

14. Dicots include _____, _____, _____, and _____. The veins in their leaves are _____, and their flower parts come in _____ or multiples of 4 or 5.

## Draw in a Sketch

15. Go outside and sketch a plant in your yard or neighborhood, considering God's creation and His grace noted in Genesis 1:11.

## Laboratory 23: Protistans and Fungi

**Lab Notes:**

### REQUIRED MATERIALS

- ☐ Microscope (from the supply kit)
- ☐ Prepared slide of *Amoeba* (from the supply kit)
- ☐ Prepared slide of diatoms (from the supply kit)
- ☐ Prepared slide of *Euglena* (from the supply kit)
- ☐ Prepared slide of *Volvox* (from the supply kit)
- ☐ Sterile toothpicks (from grocery store)
- ☐ Yeast soaked in red food coloring (from grocery store)
- ☐ Mold (grown at home as prepared last week)
- ☐ Clean blank microscope slides (from the supply kit)

### INTRODUCTION

The organisms studied in the lab exercise are different than those studied in the text, to provide greater coverage of life forms and to leave more time for the animals in labs 24 and 25.

Living organisms are broken down into the prokaryotes (without nuclear membranes and membrane-bound organelles) and eukaryotes (with membranes around their nuclei and organelles). The prokaryotes include the **Domain Archaea** and the **Domain Bacteria**. The archaea (single celled resembling bacteria) are able to tolerate extreme conditions that are fatal to other cells. Bacteria are a very diverse group of single-celled life forms. The **Domain Eukarya** includes the **Kingdom Protista**, **Kingdom Fungi**, **Kingdom Plantae**, and the **Kingdom Animalia**.

The kingdom Protista includes algae (single-celled photosynthetic forms), protozoa (single-celled non-photosynthetic forms that have to feed on something else), and seaweed which are multicellular but very different from plants. The kingdom Fungi are **saprophytes**, meaning that they have external digestion. They secrete enzymes that digest the food they come in contact with before they absorb it. If you did this, you would place your hands and arms into your food, secrete enzymes into your food and absorb the broken-down nutrients. You would not need a digestive tract — no esophagus, no stomach, and no intestines. You would not produce any solid waste. This would cut down the number of trips to the bathroom. Why do you suppose God did not make us that way? You could smell but never taste.

Diatoms are the most abundant algae found in fresh water and marine aquatic environments. They are recognized by the outer shell they secrete. These are single-celled forms and their shells have very intricate designs. This is an extreme case of the miniaturization of very detailed designs. Some secrete shells of silicon dioxide (glass) and some secrete shells of calcium carbonate (chalk).

*Euglena* is a flagellate algae, meaning that it has a long whip-like tail called a flagellum that it uses to move itself. This form feeds while in dark and uses photosynthesis in light. Botanists called them algae and zoologists called them protozoa until they got together and called them protistans. This represents the complexity of God's design rather than evolutionary intermediates. Sometimes God's design exceeds our classification abilities.

*Volvox* is a colonial algae. A colony is a group of cells joined together to contribute to each other. In a colony, the cells still carry out their individual functions; whereas, in a multicellular organism, no one cell carries out all of the life functions. Some present *Volvox* as an evolutionary stage between single-celled life forms and multicellular life forms. But *Volvox* is a colony and not a multicellular organism.

The *Spirogyra* observed in an earlier lab is another example of a single-celled algae.

The *Amoeba* is a classic single-celled protozoan. It constantly changes shape. it can fold itself around a small food object and secrete enzymes to digest and absorb it. It moves by stretching out part of itself and bringing up the rest of the cell to catch up with it. This is called amoeboid motion. Our white blood cells can do this as they maneuver through our blood stream, in and out of blood vessels, and between cells. Some of the white blood cells can engulf and digest cancerous cells, damaged cells, viruses, and bacteria.

The *Paramecium* observed in an earlier lab is another example of a single-celled protozoan.

## PURPOSE

The purpose of this exercise is to provide further experience in using the microscope and to study living and preserved members of the kingdoms Protista and Fungi.

## PROCEDURE

1. Place the prepared microscope slide of *Amoeba* on the stage of the microscope and examine it under high power. Identify its outer **plasma membrane**, **nucleus**, and **cytoplasm**. If you see part of it sticking out to the side, this is a **pseudopod** (false foot). The structures of the *Amoeba* are stained to make them easier to be seen. Point out the

*Amoeba*, its structures, and how it moves and digests food to your teacher. Perhaps you can find a video of amoeboid motion on the Internet.

2. Place the prepared microscope slide of diatoms on the stage of the microscope and examine them with high power. Spend some time observing their many diverse forms. They are also stained, making it easier to see their details. Each different structure and pattern of lines on the shell indicates a different species. Point out many of the diverse forms to your teacher. As photosynthetic cells, diatoms produce sugar (glucose) and $O_2$. They also are a major food source for many protozoa and aquatic filter feeders.

3. Place the prepared microscope slide of *Euglena* on the stage of the microscope and examine it with high power. You may not see the flagellum because it is very thin. You should be able to see a dark spot on one narrow end. This is an **eyespot** that can detect light so that it can turn itself to take advantage of maximum light for photosynthesis. It has been said that God is in the details. Point out a *Euglena* cell, its outer **cell wall** (because it is plant like), **nucleus**, **chloroplasts** (green), and **eyespot**, and their functions to your teacher. These are a very important source of $O_2$ in aquatic environments and a food source for protozoa and filter feeders.

4. Place the prepared microscope slide of *Volvox* on the stage of the microscope and examine it with medium and high power. *Volvox* are unusual in that they form colonies. They usually appear as a large sphere with partitions between its segments. Each of the segments that you observe is a separate cell. The colony is for purposes of reproduction, feeding, and protection. They are a colony rather than a multicellular organism because at stages in its life cycle, the cells will exist separately. Your body does not come apart as separate cells as a normal part of your life cycle. Point out a *Volvox* colony and its segments to your teacher. Describe its role as a photosynthetic algae.

5. Look up information and write a short paragraph on the structure and purpose (as a decomposer) of mold. Place a drop of water on a clean blank microscope slide. Use a toothpick to take a very small sample of the mold you prepared from last week and place it in the drop of water and mix it around, and without pressing down, place a cover slip on it. Do this once with the mold from bread and again with the mold growing on an orange. Examine it closely under the low power of the microscope. There may be too much material on your slide, in which case you will need to redo it with less mold and spread it out more. Point out the structures that you observe to your teacher.

6. Look up information and write a short paragraph on the structure and function of yeast. Why is it a fungus? Place a drop of water on a clean blank microscope slide. Add a very small amount of yeast soaked in red food coloring to the drop of water. Place a coverslip on it without pressing down. Examine it closely under the low power of the microscope. There may be too much material on your slide, in which case you will need to redo it with less mold (yeast) and spread it out more. The yeast should appear as small round objects. Point these out to your teacher and describe the purpose of yeast.

7. Your lab grade is based upon your paragraphs and how well you "teach" your teacher.

**Laboratory Report** (20 points possible)

Identify and describe each of the following on the microscope slides to your teacher.

*Amoeba*

*Euglena*

*Volvox*

Write a paragraph describing the structure and purpose of mold.

Write a paragraph describing the structure and purpose of yeast.

Why is yeast a fungus?

Identify and describe your mold sample and yeast to your teacher.

## Short Answer

1. What did some claim about plants and the Flood, and what did studies actually show?

## Fill in the Blank

2. _____ animals have different cells carrying out different roles.

3. In a colony, the cells can handle all of the necessary _____ when the cells are not assembled together.

4. When classifications were based upon the _____ and _____ between organisms, they were more consistent and less confusing.

5. _____ _____ and _____ _____ were creationists who contributed a great deal in the early years of taxonomy because they believed that there was _____ in creation.

6. Traditionally, _____ _____ was used almost exclusively in classifying organisms.

7. Today, _____ comparisons and _____ _____ sequences in proteins are also used to see similarities and differences.

8. A _____ is a recognized classification group.

9. The genus name *Canis* is a _____, but the name "dog" is not.

10. It is important to use _____ as well as living organisms in classification.

11. _____ is a pattern of evolutionary classification where only selected characteristics are used.

## Matching

A. Cnidaria        B. Porifera        C. Reproduce        D. Baraminology

12. _____ This is the study of trying to group organisms in their kind of creation.

13. _____ Interbreeding species are considered to be in the same kind because they can interbreed and produce offspring that can do this.

14. _____ Sponges are in this phylum.

15. _____ Jellyfish and sea anemones are in this phylum.

## Short Answer

1. What happens if an organism uses more energy in getting their food than they get out of their food?

## Matching

A. Soils          B. Leech          C. Platyhelminthes          D. Annelida          E. Nematoda

2. \_\_\_\_\_ Flatworms are in this phylum.

3. \_\_\_\_\_ Roundworms are in this phylum.

4. \_\_\_\_\_ Nematodes (a common name for roundworms) are found here.

5. \_\_\_\_\_ Segmented worms are in this phylum.

6. \_\_\_\_\_ A parasitic annelid is this.

## Fill in the Blank

7. Octopi and clams are in the phylum _____. They are called _____ bodied because they have a fleshy _____ that covers their major organs.

8. Insects are in the phylum _____. This means _____ _____ because their limbs are jointed. Their skeleton is on the outside of their body, called an _____.

9. The major classes of arthropods are the _____, _____, and _____. Two other classes include _____ and _____.

10. Arachnids include _____, _____, _____, _____ _____, and _____. They have _____ legs and _____ main body divisions.

11. Crustaceans include _____, _____, _____, _____, and _____. They have claws called _____ and _____ pairs of antennae.

12. Insects make up the _____ group in the animal kingdom. They have _____ legs and their bodies are divided into _____ parts.

13. Animals in the phylum _____ have a _____ _____ system that pumps _____ _____ instead of blood through their bodies.

14. If a sea star (starfish) loses an arm, it can _____ a new one.

15. Sea squirts and Amphioxus are _____, meaning that they do not have
_____. They are in the phylum _____, which includes animals that
have a _____ sometime in their life.

## Laboratory 24: Invertebrate Animals

**Lab Notes:**

### REQUIRED MATERIALS

- ☐ Microscope (from the supply kit)
- ☐ Prepared slide of earthworm cross section (from the supply kit)
- ☐ Prepared slide of *Planaria* (from the supply kit)
- ☐ Nitrile gloves (from the supply kit)
- ☐ Scissors with fine sharp tip (from the supply kit)
- ☐ Styrofoam dissection tray (from the supply kit)
- ☐ Safety scalpel #11 (from the supply kit)
- ☐ T-pins (from the supply kit)
- ☐ Earthworm dissection guide (from the supply kit)
- ☐ Earthworm specimen (from the supply kit)
- ☐ Grasshopper dissection guide (from the supply kit) — **Optional**
- ☐ Grasshopper specimen (from the supply kit) — **Optional**

### INTRODUCTION

This is the beginning of your dissection experience in this study. For some, the first dissection can be a bit emotional. Cutting into something that was alive can be traumatic. The reason dissections are done is to understand what the living is like. It is very different to look at the functioning and moving organs of a living animal but a bit inconvenient for the animal. This is where MRI techniques have been such a blessing. This way, your internal organs can be seen as they move and function without being invasive.

What is a biology class without dissecting an earthworm? By the way, some people eat earthworms, but I do not recommend eating the one you are dissecting. An earthworm represents the great diversity that God has created in animals. Second to insects, earthworms may be the most abundant animals on earth. They burrow through soil digesting food materials and releasing the nutrients into the soil for plant growth. As you look at the internal digestive tract of the earthworm you will see how that works.

If you run your fingers along the outer surface of the earthworm, you notice some short, pointed bristles extending down the underside of the body. These are the setae. They are used to grab the surface an earthworm is crawling along. They indicate to you which is the underside (ventral surface). **Ventral** refers to the belly side and **dorsal** refers to the back side.

The earthworm's body is divided into segments. Some of the organs are repeated in each segment. There is a large structure that looks like a collar around the earthworm called the **clitellum**. It is important for reproduction. It is also used to identify the front of an earthworm. The distance from the clitellum to the head is shorter than the distance from the clitellum to the anus.

Internally, an earthworm has a mouth, pharynx (like the back of your mouth), esophagus, crop, gizzard, and intestine. The digestive tract runs the length of its body because it takes quite a bit of time to digest food particles from soil. Surrounding the esophagus are 8 hearts (also called aortic arches) that move blood from the ventral blood vessel to a dorsal blood vessel.

## Optional Enrichment Exercise

Grasshoppers are good starting specimens. They are good representatives of invertebrate animals and they usually are not difficult. As insects, grasshoppers have an **exoskeleton** — hard outer covering to which muscles attach to move the body. They have an **open circulatory system** whereby blood flows out from the heart and leaves the arteries to ooze out through the tissues. As muscles contract and squeeze the body, blood is squeezed back to the heart rather than traveling through veins. For arthropods this is a more efficient system. The tissues are well supplied with blood while the blood travels a short distance to get back to the heart.

Along the lower portion of the abdomen of the grasshopper are tiny openings. These are the **spiracles** where air containing oxygen is carried by tiny tubes into the tissues of the body without going into lungs. For insects this is a very efficient system. As the muscles of the abdomen contract and relax, air moves into and out of the spiracles.

Insect mouthparts are very complex and varied. A butterfly tongue is like a long tube that can extend to the base of a flower and suck up the sweet nectar. The mouthparts of a grasshopper are designed for biting and chewing plant leaves.

Internally, grasshoppers have a large crop that stores eaten plant materials as it begins digestion. This is followed by a short midgut and colon. Around the outside of where the midgut and colon meet, there are a number of narrow tubes called Malpighian tubes that serve as insect kidneys.

In some cultures, people eat grasshoppers. Under dire conditions, some have had to resort to eating them. Cats love grasshoppers so if you have a problem with grasshoppers — get a cat.

## PURPOSE

The purpose of this exercise is to observe a whole mount slide of a planaria as a representative of the class Platyhelminthes; to gain dissection experience and observe the external and internal anatomy of an earthworm as a representative invertebrate.

## PROCEDURE

### Part 1

1. Place the prepared microscope slide of planaria on the stage of the microscope and examine it with low power. Planaria are in the Phylum Platyhelminthes (aka flatworms). Flatworms are as thin as a thin sheet of paper. The diatoms studied earlier had miniaturized designs on their shells and the flatworms have all of their organs in a very narrow space. Remarkable! At the front of the animal there are two projections sticking out sideways. Between them are what look like 2 eyes that are cross eyed. The internal organs are behind them. They are stained so you can see them clearly. Planaria live in fresh water streams and rivers on the surface of submerged rocks and plants. They have a narrow tube with a mouth on the end that acts like a vacuum cleaner to suck up algae and protozoa. Planaria are the exception among flatworms because they are not parasites. Most flatworms are terrible parasites that cause great damage and death to their hosts. Do some research and point out and describe the planaria and flatworms in general to your teacher.

2. Place the prepared microscope slide of the earthworm cross section on the stage of the microscope and examine it with low power. Identify the **outer epidermis**, **muscle layers** just under the epidermis, body cavity (**coelom**) outside of the intestine, **intestine**, **typhlosole**, and the cavity (**lumen**) inside the intestine to your teacher.

### Part 2

1. Lay out the Styrofoam dissection tray along with the scissors, scalpel, and pins. Be careful that you do not cut into the Styrofoam while doing the dissection. It will come in handy because you will need to pin down the specimen to the Styrofoam.

2. Put on the gloves and examine the external anatomy of the earthworm. Use the earthworm dissection guide to help you find the structures. Find the **clitellum** which is a thick collar-shaped structure. The shorter section from the clitellum is the anterior portion. The **mouth** is at the end of this section. The opposite end at the end of the long section is the **anus**. Run your fingers down the length of the

earthworm and feel the scratchy bristles called the **setae**. They are on the **ventral** (belly) side of the worm. The back side is the **dorsal** side. Place the worm on the tray with the ventral side down.

3. The mouth is the opening of the first segment (looks like a knob on the front) called the **prostomium**.

4. Pin the prostomium down to the Styrofoam with a pin. Pull the worm out lengthwise and pin the end with the anus down. With the scalpel, carefully cut along the dorsal side from the prostomium back to the clitellum. Do not cut very deep. The internal segments are separated by membranes called septa (plural of septum). Gently cut the septa on the right and the left of the internal organs and pin down the sides of the worm.

5. You can see the digestive tract from the mouth back. Count back to segments 4 and 5. You can see an expanded section of the digestive tract called the **pharynx**. The muscular lining of the pharynx draws soil into the mouth and into the pharynx. After the pharynx is a narrower **esophagus** that carries soil to the larger **crop** (at about segment 15). This is followed by the **gizzard** which together with the crop grinds up the soil and mixes enzymes with it to break down the food material in the soil. After the gizzard is the **intestine** where further digestion and absorption of nutrients occur. The intestine looks like a flat tube running the length of the earthworm to the anus where soil and broken down organic material is expelled. The expelled soil is rich in nutrients for plants. Through this process, earthworms loosen the soil to aid air penetration and add fertilizer, making it an ideal habitat for plant roots.

6. Go back and look carefully ahead of and behind segment 10 and you should see 4 sets of narrow tubes coming around the esophagus on the right and left sides. These are 8 (4 sets of 2) **aortic arches** or **hearts**. They carry blood from the ventral artery that runs along the inside of the ventral side of the body cavity to the dorsal artery that runs along the inside of the dorsal surface of the body cavity. You will also see a number of white structures in this region of the body cavity — these are the reproductive organs. Earthworms are hermaphrodites, meaning that each individual worm is both male and female. They join at their clitella and secrete eggs and sperm cells into a cocoon formed around their bodies. The egg and sperm cells join nuclei and form embryos. This gives the embryos a new combination of chromosomes that means that they will have fewer double (homozygous) recessive harmful genes.

7. Describe in detail your procedure with the earthworm to your teacher. Point out all of the structures in bold print and describe the function of each. It is not enough to know the names of structures in anatomy: their functions are also very important. Your lab grade is based upon how well you do the dissection and explain your findings. Afterward, wrap the earthworm in a paper towel and dispose of it in the trash. After you are done, wipe off the surface you worked on with a dilute bleach solution.

**Part 3: Optional Enrichment Exercise**

1. Take the grasshopper and examine its external anatomy. Use the dissection guide to help find the following structures.

   A. **Head**
   B. **Thorax** (chest)
   C. **Tympanum** (eardrum — this is just above the hind leg)
   D. **Abdomen**
   E. **Antenna** (used to feel things nearby)
   F. **Compound eye** (each compound eye is composed of many smaller eyes)
   G. **Femur**
   H. **Tibia**
   I. **Spiracles** (tiny holes along the lower side of the abdomen)
   J. **Forewing**
   K. **Hindwing** (the forewing rests on top of the hindwing)
   L. When the posterior of the abdomen comes to one point it is a male. When it comes to an upper and a lower point it is a female and they are called **ovipositors**.

2. Using the scissors, make a cut along the lower part of the abdomen. Cut up from the front and back of this cut to the top so that you can raise the covering to expose the inner part of the abdomen. The covering will be attached to the underlying tissue, so you will need to cut it loose.

3. Identify the following internal structures within the abdomen.

   A. **Crop** (the anterior part of the digestive track)
   B. **Midgut** (second part of the digestive track)
   C. **Gastric ceca** (finger-shaped pouches that come out from between the crop and the midgut, these can expand to hold partially digested food when it comes through too fast)
   D. **Colon** (the posterior part of the digestive tract, completes digestion and eliminates waste)
   E. **Malpighian tubules** (stringy-looking strand that extends anterior and posterior from where the gastric

ceca and the colon meet; these are the kidneys of the grasshopper)

F. Along the inner top surface of the abdomen of the grasshopper is a thin membrane that may look like a thin tube. This is the **heart**. There are openings along the lower surface in the middle of the heart called **ostia**. This is where blood re-enters the heart. The heart of ostia are more difficult to see. You can cut into these if you are not careful. This is part of the open circulatory system of arthropods. This may be hard to see, so do not worry if you cannot be sure.

4. Describe in detail your procedure with the grasshopper to your teacher. Point out all of the structures in bold print and describe the function of each. It is not enough to know the names of structures in anatomy: their functions are also very important. Your lab grade is based upon how well you do the dissection and explain your findings. Afterward, wrap the grasshopper in a paper towel and dispose of it in the trash. After you are done, wipe off the surface you worked on with a dilute bleach solution.

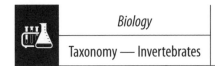
**Laboratory Report** (20 points possible)

Describe the procedure, structures, and their functions of your earthworm dissection to your teacher.

Describe the procedure, structures, and their functions of your grasshopper dissection to your teacher.

## Fill in the Blank

1. Vertebrates are in the phylum _____ and subphylum _____ (or
   _____).

2. Chordates have a rod of cartilage called a _____ and a _____
   _____ nerve cord that coordinates nerve impulses from the brain to the rest of the body.

3. Some vertebrates have the _____ only as embryos. The vertebrae develop around it.

4. The _____ is the anterior portion of the digestive tract. As embryos, chordates have
   _____ _____ extending out from the sides of the pharynx.

5. These pouches form _____ in fish and tadpoles. They form _____
   _____, and support for the _____ and _____
   _____ in other vertebrates.

6. Chordates (except humans) have a _____ behind the anus.

7. Fish without jaws for biting and chewing (they have round mouths) are in the class
   _____. Extinct fossil jawless fish are _____, which had dermal
   _____ _____ on their heads and bodies.

8. _____ enter _____ _____ through their gills and
   anus and eat them from the inside out.

9. _____ attach with their round suction cup mouths to the sides of
   _____ _____ and eat them from the outside.

10. Fish with skeletons made of cartilage are in the class _____. These include
    _____, _____, and _____. Their only bones are in
    their _____ and _____.

11. Fish have numerous methods of reproduction. _____ fish embryos develop outside of
    their mother's body. _____ fish embryos develop within their mother's body without
    drawing nourishment from her body. _____ fish embryos develop within their mother's
    body and get nourishment from her body.

12. Sharks have multiple rows of _____ that replace each other.

13. Bony fish are in the class _____. Their gills are covered by a bony covering called an
    _____.

14. _____ remove oxygen from water and _____ remove oxygen from air.

## Fill in the Diagram

15. Label the diagram of the fish circulation pattern. (page 269)

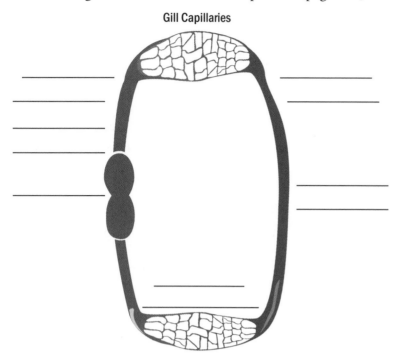

**Gill Capillaries**

16. Label the diagram of the land vertebrate circulation pattern. (page 269)

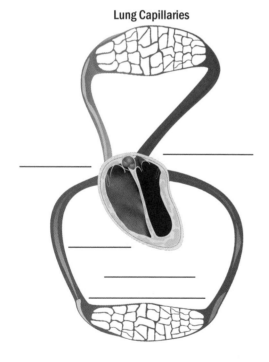

**Lung Capillaries**

## Matching

A. Urodela          B. Amphibia          C. Apoda

17. _____ Vertebrates in this class are land animals that have to stay moist.

18. _____ Caecilians are amphibians with sharp teeth and no legs, and are in this order.

19. _____ Salamanders are in this order.

*Biology*

Taxonomy — Vertebrates

Pages 271–277

Day 152

Lesson 25
Worksheet 2

Name

## Short Answer

1.  Why do some feel reptiles should be divided into several classes?

## Fill in the Blank

2.  Frogs and toads are in the order _____, which means _____.

3.  Reptiles are _____, meaning that they get most of their body heat from their _____. They are also _____, meaning that they can survive with a wide range of _____ _____.

4.  The order _____ includes tortoises and turtles.

5.  Lizards and snakes are in the order _____.

6.  Crocodiles and alligators are in the order _____.

7.  "Lizard-hipped" dinosaurs are in the order _____.

8.  "Bird-hipped" dinosaurs are in the order _____.

9.  Birds are in the class _____. Birds have _____ bones that enable them to fly by lightening and strengthening the bones.

## Short Answer

10.  Contrast the initial response to the fossil *Archeopteryx* with the currently accepted view.

**Matching**

   A.  Monotremes      B.  Eutheria      C.  Mammalia      D.  Metatheria      E.  Theria

11. _____ Mammals are in this class.

12. _____ The subclass Prototheria are egg-laying mammals called this.

13. _____ This subclass includes mammals whose embryos develop within their mother's uterus and are nourished through a placenta.

14. _____ Embryos of mammals in this infraclass continue their development outside the uterus in a pouch where they feed on mammary glands.

15. _____ Embryos of mammals that develop within their mother's body right up to birth are in this infraclass.

## Laboratory 25: Vertebrate Animals

**Lab Notes:**

### REQUIRED MATERIALS

- ☐ Microscope (from the supply kit)
- ☐ Prepared slide of frog ovary (from the supply kit)
- ☐ Prepared slide of frog sperm (from the supply kit)
- ☐ Prepared slide of human skin (from the supply kit)
- ☐ Nitrile gloves (from supply kit or locally)
- ☐ Scissors with fine sharp tip (from the supply kit)
- ☐ Styrofoam dissection tray (from the supply kit)
- ☐ Safety scalpel #11 (from the supply kit)
- ☐ T-pins (from the supply kit)
- ☐ Frog dissection guide (from the supply kit)
- ☐ Frog specimen (from the supply kit)

### INTRODUCTION

Vertebrates are very different than invertebrates. Usually, the first vertebrate dissection experience is with a frog. They are easier to dissect, and provide a good introduction to vertebrate anatomy.

Histology (study of tissues) is an important study of anatomy. The anatomy of the frog's reproduction system is studied by observing the cross section of a frog ovary that has eggs at different stages of development. Frogs release their eggs all at once in a pond or similar still body of water. The male frog releases a fluid containing sperm cells into the water over the eggs. Zygotes form and develop into tadpoles which later develop into mature frogs. This pattern of development enables amphibians to live in moist habitats where many other organisms could not. Where would you have to live if you developed into a tadpole in the early part of your life? Your possible habitats would be much more restrictive than they are now. God knew what He was doing when he created all living organisms for their particular habitats.

The histology of human skin is included in this study of vertebrate structures. A frog's skin is much thinner than ours. They absorb oxygen through their skin and we do not. This is one of the reasons amphibians have to be in a moist environment. If their skin dries out, their skin no longer absorbs oxygen and they suffocate.

Biology is not the study of dead things. It is a study of the living, but many times a dead animal is studied to see what the living is like. For example, if a surgeon needs to practice before performing a surgery, a practice surgery will be performed on a cadaver (dead body) first. Then the real surgery should go well. I am sure that as a patient any of us would appreciate that. For some of you, this will be your first dissection of a vertebrate. The preservative used is safe, but do not nibble on it. Yuck! Wear gloves. You may not realize it now, but some of you doing this study will go on to become nurses and doctors. If it makes you nervous — that is okay. Take your time. The dissection guide is to be a reference to help you along with the instructions given below.

The point of the dissection is to learn what the normal structures look like and where they are. Therefore, do not remove things. Some will cut out structures and lay them on a piece of paper with labels so that they can identify them. That is **not** the purpose of this dissection. The purpose is to learn what the structures look like and where they belong. A good dissection leaves the structures in place. That is easier to do with a frog, which is another reason for using them.

As you go through this exercise you may need to go back to some previous labs and chapters to review the meanings of some terms.

## PURPOSE

The purpose of this exercise is to study vertebrate histology and proper dissection techniques, focusing on the names of anatomical structures, what they look like, and where they are in the intact organism. Even though the frog is dead, the purpose is to learn the anatomy of a live frog without causing pain to a live frog. As well, this exercise may provide the beginning of what may become a profitable career in research and medicine for some of you. Some curricula use virtual dissections. They can be helpful to explore the anatomy of other vertebrae forms, but it is still necessary to work with an actual specimen. It is important to remember that all life was created by God and is valuable to Him. This means that it is never right to be cruel or cause unnecessary pain to an animal.

## PROCEDURE

1.  Place the prepared microscope slide of the cross section of a frog ovary on the stage of the microscope and examine it with low power. The ovary is divided into the outer region called the **cortex** and the inner region called the **medulla.** The developing **eggs** are in the outer cortex. The eggs are unusually large cells because they have to have enough cytoplasm to provide food for the developing embryo before it can feed as a tadpole. When the eggs mature and get larger, they cause the outer **epidermis** to appear very lumpy. Point out the ovary and the structures to your teacher.

2. Place the prepared microscope slide of the frog sperm cells on the stage of the microscope and examine them with high power. The sperm cells are quite small because all they contribute to the developing zygote is their nuclei. They have a main **cell body** (that contains the nucleus) and a long tail-like **flagellum** with which they swim around to contact the eggs after being released into a pond or stream. Point out the sperm cells and their structures to your teacher.

3. Place the prepared microscope slide of the cross section of human skin on the stage of the microscope and examine it with low power. This slide is to expand your knowledge of vertebrate histology. The top portion of skin is the **epidermis** and the lower portion is the **dermis**. The dermis is not as compact as the epidermis. The prefix epi- means above, so the epidermis is above the dermis. The dermis contains nerve cells, blood vessels, adipose (lipids) tissue, and fluid. Look for **hair follicles** from which hair shafts grow. You may see some **oil glands** that appear as a bulb-shaped structure in the dermis with a tube that carries oil to the surface of the epidermis. Point out the structures that you identified to your teacher.

4. Lay out the Styrofoam dissection tray along with the scissors, scalpel, and pins. Be careful that you do not cut into the Styrofoam while doing the dissection.

5. Lay the frog on the dissection tray and using the diagram below and the frog dissection guide, locate the following external structures. As you go through this exercise you may find it helpful to search out additional helps with pictures of a frog dissection. After you identify these structures you will have to point them out and name them for your teacher without using any helps — so learn them as you are identifying them.

   A. External nares
   B. Eye
   C. Nictitating membrane (this is a membrane that can cover the eye and keep it moist)
   D. Tympanum (this is the frog's eardrum, which is just behind the eye)
   E. Foot
   F. Forelimb
   G. Hindlimb

6. Lay the frog on its back on the tray. Make a snip on the lower abdomen and cut up toward the region between the forelimbs. Make side cuts to the right and left of the first cut at the top and bottom of the first cut.

7. Lay the skin back to expose the internal cavity (coelom). The frog is easier to work on because it does not have muscles just under the skin as in mammals.

8. Using the dissection guide and the diagram below, locate the following organs. You could also find some good photographs of a frog dissection on the internet — just search "frog dissection."

   A. Coelom (body cavity)
   B. Pericardium (the membrane covering the heart, which is the middle of the chest)
   C. Heart
   D. Lung
   E. Liver
   F. Gall bladder (stores bile salts from the liver that aid in digestion)
   G. Stomach
   H. Intestine
   I. Fat bodies
   J. Pancreas (in the fold where the small intestine connects to the stomach, produces additional bile salts and insulin to aid cells in absorbing glucose [sugar])
   K. Kidneys
   L. Spleen (stores red blood cells in case of bleeding)
   M. Ovaries (females, produces egg cells by meiosis) You might see some eggs in the ovaries.
   N. Oviduct (females, carries eggs from the ovaries)
   O. Testes (males, produces sperm cells by meiosis)
   P. Vasa efferentia (carries sperm cells from the testes)

9. Describe in detail your frog dissection procedure to your teacher. Point out all of the structures and briefly describe the function of each. It is not enough to know the names of structures in anatomy; their functions are also very important. Your lab grade is based upon how well you do the dissection and explain your findings. Afterward, wrap the frog in a paper towel and dispose of it in the trash. After you are done, wipe off your work surface with a paper towel and a dilute bleach solution.

## Laboratory Report (20 points possible)

Describe your procedure, structures, and their functions of your frog dissection to your teacher.

## Short Answer

1. How does the text describe the world in the beginning, and all that happened through the Flood?

## Fill in the Blank

2. God created all living organisms after their _____.

3. Microevolution refers to _____ changes in already _____

   _____.

4. _____ has been used to refer to the origin of life itself.

5. The guidance of the _____ _____ makes a huge difference in how life can be interpreted. To believe in creation, you have to have a _____ who is outside of the creation.

6. In 70 B.C. Lucretius said that the _____ itself yielded plant and animal forms.

7. Prior to the 1700s, most geologists and mining engineers believed that the fossil-containing rock layers were formed by the _____ of _____.

8. Later, James Hutton proposed the idea called _____, which meant that everything was formed by natural processes and not catastrophes or miracles.

## Matching

A. Fossils     B. Pangenesis     C. Helens     D. Environment     E. Lyell

9. _____ Modern geologists have come to realize that catastrophes can cause major sudden changes as seen in the eruption of Mount St.:

10. _____ William Smith said that rock layers could be dated by the types of what that they contained?

11. _____ This Charles wrote the first major textbook in geology.

12. _____ Jean Baptist de Lamarck presented a philosophy called:

13. _____ Natural selection means that some organisms survive better than others in their natural:

## Fill in the Blank

1. Darwin wrote a book titled _____
   in which he stated that all the diverse life forms have come about from very _____
   _____ in each generation over _____ of years.

2. _____ fossil forms between kinds of creation have _____ been found.

3. In the early _____ many became critical of Darwinism because of modern Mendelian
   _____ and studies in _____. This time period was called the
   _____ _____ because if evolution were not true, they did not know
   where life came from.

4. In trying to explain the lack of intermediate fossils, Eldredge and Gould proposed a concept called
   _____ _____ where evolution occurred in rapid events.

5. _____ rocks are formed by the accumulation of soil from wind or water erosion. Fossils
   are best buried _____ rather than _____.

6. _____ rock is formed by the cooling of molten lava. These do not contain
   _____ because the melted rock would have destroyed them.

7. _____ rock is formed by massive _____ and high
   _____. These also do not contain _____.

## Short Answer

8. Contrast the geological column with the global Flood.

## Matching

|   | A. Day Age | B. Theistic evolution | C. Literal historical | D. Gap |

9. _____ In trying to stay true to Genesis, many adopted this theory, sometimes called the Ruin and
   Reconstructionist Theory.

10. _____ More recently, this theory became popular, stating that the days of creation were actually long
    periods of time.

11. _____ These people say that macroevolution occurred and all the problems with it are answered by
    saying that God did it.

12. _____ The Hebrew language in the Genesis accounts of creation and the Flood are written in this style.

## Short Answer

13. Address the features of the young earth creation view based on the biblical record.

## Laboratory 26: Decomposition or Fossilization

### REQUIRED MATERIALS

- ☐ Notepad, pencil, phone or camera to take pictures
- ☐ A partially eaten chicken leg or similar piece of meat — **Optional**
- ☐ A piece of screen or chicken wire about 1½ feet x 1½ feet — **Optional**
- ☐ A plot of ground 1 foot x 1 foot and 2 inches deep — **Optional**
- ☐ Wet soil — **Optional**

**Note:** You may want to reference Lab 10, which utilized these same materials.

### INTRODUCTION

Materials that are buried gradually without being protected do not usually form fossils, but are carried off and eaten by critters. To form a fossil, dead tissue has to be rapidly buried deep enough, or protected if it is gradually buried. This is cited by the theory of punctuated equilibrium to account for the lack of fossils showing evolution from one form to another. Most fossils are hard tissue, such as bone, and tissues they contain.

Some fossils form when a plant or animal dies and is deposited onto soft mud. The organism makes an impression in the mud. Then the organism decomposes, and the mud hardens with the impression of the organism. Some common examples are fish skeletons and plant leaves.

Most fossils are found clumped together in large masses with evidence of rapid burial and water erosion from flooding. Charles Darwin said that if fossils representing forms intermediate between major taxonomic groups are not found in the next 100 years, his idea of gradual changes over long ages would be false. More than 100 years have gone by and those intermediates still have not been found. To phrase it from a creationist perspective, fossils fall into distinct groups that represent the kinds of creation.

The major threat to fossil formation with gradual burial is animals carrying it off as food or the process of decomposition. Besides destroying potential fossils, decomposition is a major step in recycling nutrients back into food webs. Animal waste and animal and plant bodies are broken down into their basic molecules and become nutrients for producers (photosynthetic forms) to produce new carbohydrates, proteins, lipids, and

**Lab Notes:**

nucleic acids. Major agents of decomposition include bacteria, fungi, and fire.

## PURPOSE

The purpose of this exercise is to observe decomposition processes that work against fossilization.

The optional exercise that follows demonstrates how fossils form and what is helpful to the process and what works against the process.

## PROCEDURE

1. Take a hike in a pleasant outdoor setting and look for animal or plant waste — dog droppings, leaf litter, parts of insects or insect cocoons, etc. Often you can find clumps of fallen leaves around trees or caught in bushes. Look under the leaves where moisture is trapped away from sunlight. Also check under rocks where moisture is retained. Describe what you observe and document it with pictures. These represent decomposition by bacteria.

2. Check for mold. Describe and take pictures of what you observe. This is where fungi are the active agents breaking down dead matter.

3. Look for lichen growing on rocks. Describe and take pictures of what you observe. Lichen is a combination of algae cells growing between the cells of a fungus. The algae provide glucose and $O_2$ for the fungus and the fungus supplies $CO_2$ and water for the algae. This is a symbiotic relationship called mutualism. The lichen is breaking down rock into fine sand, providing a habitat for plants to take hold and grow.

### Optional Enrichment Exercise

Back in lab 10 you may have prepared a fossil to study in this exercise. The big day has arrived. Dig up your bone, take a picture of it and describe it in detail. Read over the journal that you kept since that time. Are there any insects around the bone or that general area? Describe in detail what happened to the fossil over these past weeks. You can break it down into weeks or groups of weeks. From your observations answer these questions. Also describe how your experiment was used to answer each question.

1. Which is most likely to form a fossil — rapid burial or very slow burial where soil is deposited very little at a time?

2. What happens to a fossil in the ground?

3. How is a fossil different from the original bone structure?

4. Is all of an animal preserved as a fossil?

5. What does the fossil tell you about the original animal?

6. How would fossils have been buried in a flood?

7. Would a fossil form as well if it had wind gradually blowing sand over it without any wire covering?

**Laboratory Report** (20 points possible)

Answer to Question 1

Answer to Question 2

Answer to Question 3

Answer to Question 4

Answer to Question 5

Answer to Question 6

Answer to Question 7

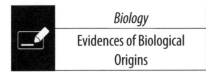
## Fill in the Blank

1. Comparative anatomy is not used to prove that macroevolution _____ but to try to find the _____ of evolution. Evolution cannot be proven to be _____ because it cannot be _____ or _____.

2. Birds and bats have similarities because they both fly. However, bats' metabolism is much _____ and they have no _____. Bats are _____, not birds.

3. In evolution theory, the wings of birds and bats are called _____ evolution, which means that similar structures evolved more than once in different organisms.

4. God created many different kinds of _____ and created _____ in the groups of organisms to fill each _____.

## Draw the Diagram

5. Draw the diagrams of the wing structure of bats and birds on page 300.

| | |
|---|---|
| | |
| Wing structure of bats | Wing structure of birds |

## Matching

    A. Sixth         B. Fifth         C. Radioisotope decay     D. mtDNA     E. Molecular

6. _____ Birds were created on this day.

7. _____ Mammals were created on this day.

8. _____ In evolution theory, this kind of clock is used to indicate relatedness and evolutionary time.

9. _____ This is compared between organisms to determine how much time it took for them to evolve.

10. _____ A lot of the ages assigned to the geological column are determined by this, though its accuracy is controversial.

## Matching

A. Vestigial organs     B. Purpose that works     C. Closely related     D. Application of data

1. _____ In the evolution theory, organisms with fewer differences in their DNA are considered more:

2. _____ The issue between creationists and evolutionists is not with the data (DNA differences) but rather the:

3. _____ This term refers to organs that had functions in earlier ancestors but have since evolved so that they no longer have a purpose.

4. _____ A major difference between evolution and creation is that evolution does not expect everything to have a purpose or to even work, but God designs things with:

## Fill in the Blank

5. A tailbone in humans is given as an example of a _____ organ, but without it, you would not be able to _____ or even _____.

6. It has been estimated that about _____ of the life forms created are still around. The rest would have gone extinct as shown by the fossils.

7. Fossils usually appear in the same _____. This convinced most that the earth was very _____ and not _____. The eruption of Mount St. Helens helped to replace opinion with _____, _____ data.

8. At first, Charles Darwin thought that _____ were isolated and could not _____. He observed in his travels that many species _____ each other and many were found in only _____ places.

9. Darwin later believed he could explain his observations by _____ causes rather than _____. If Darwin had realized that several _____ could be in the same _____ of creation, he would have seen a different picture.

## Short Answer

10. The text stated that: "Since the fossils usually appear in the same order, it seems reasonable that those that are deeper were buried before those that were buried above them. This was one of the strongest evidences in the 1800s that convinced most that the earth was very old and not young, as previously thought." What was needed to change this perception and why?

## Laboratory 27: Diversity Within the Kinds of Creation

**Lab Notes:**

### REQUIRED MATERIALS

☐ These depend upon what you have available. Be creative and adventurous.

### INTRODUCTION

As we go through life, we learn from things that we discover along the way. Our journey is one of learning and growing. We learn as individuals and we learn as a group. Not too long ago it was thought that species and kinds were the same. As creationists, that gave us a position that was very hard to defend. We saw the variation around us, and when species formed hybrids, as Darwin observed with the finches on the Galapagos Islands, many doubted the Genesis account of creation. In recent years, it was realized that many that were called separate species were interbreeding. This led to the conclusion that the issue was really in how the word *species* was assigned and that several species could be in the same kind. It was especially helpful when discoveries in modern genetics provided ways that the variations producing species could happen without evolution and as a normal process already built into the complex designed DNA of all life forms.

### PURPOSE

This exercise demonstrates the diversity in the kinds of creation and the barriers between kinds as well.

### PROCEDURE

1. Come up with as long of a list of canids (dogs that are tame and wild) as you can. Begin with a list of all of the domestic dogs that you can then do the rest of the canids. Check out your pets and the neighborhood. Go to other sources. Separate them into groups that cannot interbreed with each other. Be careful, because even though a Saint Bernard and a chihuahua cannot interbreed, what if the chihuahuas were not always that small and if Saint Bernards were not always that big? For example, in the 1600s the average humans were shorter. In chapter 21, some mechanisms for variation within kinds were introduced. These were not evolution but designed methods using enzymes coded for by designed DNA that act upon already existing DNA.

2. Do the same thing for cats — domestic and wild. For example, the mating of a tiger with a lion has been observed, which produced a liger.

3. Your results are to be written up in your lab report. Be sure to organize it so that someone else can follow it clearly. Use complete sentences and watch for spelling and grammar.

**Laboratory Report** (20 points possible)

Different Canids

Groups of Canids that Naturally Interbreed

Different Cats

Groups of Cats that Naturally Interbreed

Some that do not naturally interbreed may do so under different conditions. An example would be wolves and coyotes. Most creationists consider all of the canids to be in one kind of creation and all of the cats to be in one kind of creation.

## Short Answer

1.  Summarize the author's introduction about whether or not we are animals.

## Fill in the Chart

2.  Fill in the missing information from the classification chart about humans below.

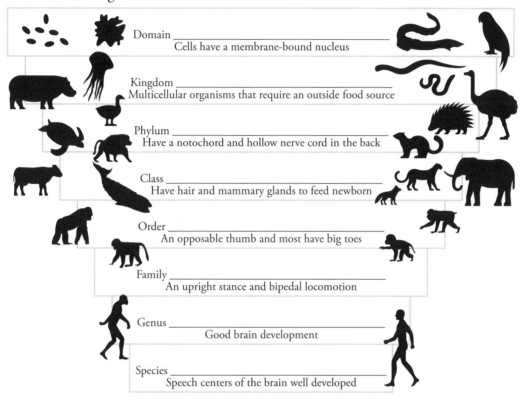

Domain _____
Cells have a membrane-bound nucleus

Kingdom _____
Multicellular organisms that require an outside food source

Phylum _____
Have a notochord and hollow nerve cord in the back

Class _____
Have hair and mammary glands to feed newborn

Order _____
An opposable thumb and most have big toes

Family _____
An upright stance and bipedal locomotion

Genus _____
Good brain development

Species _____
Speech centers of the brain well developed

## Fill in the Blank

3.  Besides being flesh, we are _____ beings created in the _____ of _____.

4.  Christ may have had _____ cells (because He had no human father) as a human, which would have made Him _____ _____ to suffer from _____ _____.

5.  We have many physical qualities that we do not _____ with other life forms. We have _____, the ability to _____ and _____, walk _____, and be creative with _____ and _____.

## Matching

A. Theistic Evolution      B. Java and Peking man      C. Day Age View      D. Gap Theory

6. _____ Consider cavemen to be pre-human forms that were destroyed in Genesis 1:2.

7. _____ Differ as to whether the cavemen were human ancestors or a different group of creatures altogether.

8. _____ Believe that cavemen were our ancestors and that there was never an actual Adam and Eve.

9. _____ Were at first thought to have been links between humans and an ancient ancestor.

## Fill in the Blank

1. _____ man turned out to be a _____. The fossil was made from part of a _____ skull and an altered _____ _____ of an _____.

2. _____-_____ man was concluded to be in the genus and species of _____ _____.

3. In 1925, _____ _____ found a skull that came to be known as _____, which means _____ _____. Its human resembling features include the size of the _____ case and the position of the opening in the base of the skull for the spinal cord, called the _____ _____.

4. *Australopithecus* _____ had a delicate anatomy, and *Australopithecus* _____ had very large molar teeth. Louis Leakey found fossils he called _____ _____, which means _____ because they found crude _____ nearby. Australopithecus _____ (nicknamed _____) was thought to have had more in common with *Homo habilis* and modern _____. All of these were later considered to be _____.

5. The basic conclusion is that we have fossils of _____ and fossils of _____, and no _____ connecting the two.

## Matching

A. Biological     B. Fossil data     C. Egg     D. Conserved     E. Eve

6. _____ Mitochondria are inherited from the mother through this.

7. _____ Wilson and Cavalli-Sforza suggested that mtDNA could be traced through female lines back to an ancestral human mother they called:

8. _____ This, when fully studied, will not contradict the Bible or our ancestry from a literal Adam and Eve.

9. _____ Because we were created in God's image, we have a spiritual nature as well as:

10. _____ There are many enzyme systems directed by common DNA in all forms, which evolutionists call these kind of traits.

## Short Answer

11. Summarize the author's last paragraph about us being both biological and spiritual.

## Laboratory 28: Human Origins

**Lab Notes:**

### REQUIRED MATERIALS

☐ Pictures of Neanderthal, Cro-Magnon, *Australopithecus afarensis* (Lucy), chimpanzee, and human skulls and skeletons as provided below.

### INTRODUCTION

When skeletal remains of the "cavemen" were first discovered they were thought to be a gold mine to show human evolution from ape-like ancestors. It is important to realize that macroevolution and theistic evolution do not have Adam and Eve being created as the first humans. There was a great deal of excitement when the first skeletal remains were found that many hoped would show the gradual transition of humans from ape-like forms. As time went by, and more data were available, it became apparent that the fossils were either human or ape-like, with no transition forms in between.

For this exercise, it would be rather difficult for you to get hold of the actual skulls and skeletons of the fossils and present-day humans, so pictures of the actual skulls and skeletons are provided. By seeing the differences and similarities for yourself, your thinking will be clearer.

### PURPOSE

This exercise will enable you to see for yourself the similarities and differences between humans, apes, and fossil forms. It also provides an opportunity to further develop your writing skills, observe and record data, and draw appropriate conclusions.

### PROCEDURE

1.  Look at the pictures of the human, chimpanzee, Neanderthal, Cro-Magnon, and *Australopithecus afarensis* (Lucy) skulls.

    A.  Describe the similarities and differences between the human and chimpanzee skulls. This gives you the information to determine if a skull is more like a human or an ape.
    B.  Look at the pictures of the Neanderthal, Cro-Magnon, and *Australopithecus afarensis* (Lucy) skulls. Label each as more like a human or more like an ape. List your reasons for each one using complete sentences.

2. Look at the pictures of the human, chimpanzee, and Neanderthal skeletons.

   A. Describe the similarities and differences between the human and chimpanzee skeletons. This gives you the information to determine if a skeleton is more like a human or an ape.

   B. Look at the picture of the Neanderthal skeleton. Label various sections as more like a human or more like an ape. List your reasons for each one using complete sentences.

3. Based upon your observations, what can you conclude about humans from the comparisons to a present-day chimpanzee (ape) and the fossil "cavemen"? Use complete sentences. Be sure that you can point to specific observations for each of your conclusions.

## Laboratory Report (20 points possible)

Similarities and differences between human and chimpanzee skulls and skeletons

For the following, indicate whether it is more human like or more ape (chimpanzee) like and why. The why is very important.

Neanderthal

Cro-Magnon

*Australopithecus afarensis* (Lucy)

What can you conclude about humans and the fossils and why? Use specific observations.

# Quizzes

**Q**

*Biology*
Cell Membranes and Nucleus

Quiz 6
Lesson 6

Day 35

Total score:
_____ of 100

Name

## Circle the Correct Answers

1.  Cell membranes consist of 2 layers of _____.

    A.  protein

    B.  DNA

    C.  phospholipids

    D.  carbohydrates

2.  Cell membranes _____, which means that only certain substances can pass into or out of the cell.

    A.  are plastic

    B.  are selectively permeable

    C.  are rigid

    D.  have large pores that are permanently open

3.  The _____ is the control center of a cell.

    A.  outer plasma membrane

    B.  proteins in the outer membrane

    C.  cytoplasm

    D.  nucleus

4.  The information to control the inheritance of physical characteristics is in _____.

    A.  DNA

    B.  the nuclear membrane

    C.  protein

    D.  phospholipids

5.  The control center of a cell was discovered using the green algae _____.

    A.  *Paramecium*

    B.  newt

    C.  *Acetabularia*

    D.  *Euglena*

## Match the Correct Answers

_____ 6.  Gene

_____ 7.  Cytoplasm

_____ 8.  Chromatin

_____ 9.  Nucleolus

_____ 10. Chromosomes

A.  contains RNA and protein

B.  section of DNA that determines physical traits of a cell

C.  DNA and its associated proteins together

D.  contain DNA with genetic code for structures of proteins

E.  material between outer cell membrane and nuclear membrane

**Circle the Correct Answers**

11. Hans Spemann found that the _____ of newt embryo cells determined what the embryo would look like.

    A. plasma membrane

    B. nucleus

    C. cytoplasm

    D. environment

12. After several cell divisions, each of the cells of a newt embryo was found by Spemann to have as much genetic _____ as the original cell.

    A. information

    B. cytoplasm

    C. protein

    D. water

13. Hammerling found that the shape of the cap of the algae he studied was determined by what was in the _____.

    A. cytoplasm

    B. cap of another cell

    C. cap of the same cell

    D. nucleus of the algae cell

14. The layers of a cell membrane have interspersed _____ molecules.

    A. protein

    B. oxygen

    C. DNA

    D. hydrogen

15. Cell membranes can _____, which is why they are called fluid and plastic.

    A. change shape

    B. disappear

    C. break

    D. pour out like water

## Circle the Correct Answers

1. By diffusion, molecules move from where they are _____ concentrated to where they are _____ concentrated.

   A. more, more

   B. less, less

   C. more, less

   D. less, more

2. When NaCl is dissolved into water, it is called a _____.

   A. solution

   B. solvent

   C. solute

   D. salient

3. Physiological saline has _____ % NaCl.

   A. 0

   B. 0.9

   C. 1.8

   D. 2.7

4. Physiological saline is _____ to the fluid within a cell.

   A. hypotonic

   B. hypertonic

   C. isotonic

   D. hair tonic

5. Distilled water is _____ to the fluid within a cell.

   A. hypotonic

   B. hypertonic

   C. isotonic

   D. hair tonic

6. 10% NaCl is _____ to the fluid within a cell.

   A. hypotonic

   B. hypertonic

   C. isotonic

   D. hair tonic

7. When a cell is placed in 10% NaCl, it will _____.

   A. remain the same

   B. become salty

   C. swell

   D. shrivel up

8. The result in question 7 occurs because _____ water diffuses into the cell than diffuses out.

   A. more

   B. the same amount of

   C. less

9. When a cell is placed in distilled water, it will _____.

   A. remain the same

   B. become salty

   C. swell up and burst

   D. shrivel up

10. _____ is where specialized proteins in the cell membrane move molecules into and out of cells.

   A. Exchange

   B. Diffusion

   C. Membrane pumps

   D. Active transport

## Match the Correct Answers

_____ 11. Solvent

_____ 12. Turgor pressure

_____ 13. Membrane pumps

_____ 14. Osmosis

_____ 15. ATP becomes ADP

A. water pressure in cells

B. some proteins in the phospholipid layers of a cell membrane

C. when energy is released from ATP

D. other molecules will dissolve into it

E. water diffusing from higher to lower concentration

## Match the Correct Answers

_____ 1. Centrioles

_____ 2. Cilia

_____ 3. Mitochondria

_____ 4. Endoplasmic reticulum

_____ 5. Ribosomes

A. where ATP is formed

B. where proteins are formed

C. small hair-like structures

D. large sheet-like organelles

E. barrel-shaped organelles in animal cells

## Circle the Correct Answers

6. _____ are an extension of the membranes of the endoplasmic reticulum.

    A. centrioles

    B. mitochondria

    C. Golgi bodies

    D. ribosomes

7. Photosynthesis occurs in _____.

    A. Golgi bodies

    B. chloroplasts

    C. mitochondria

    D. centrioles

8. The inner folds of mitochondria are called _____.

    A. cristae

    B. ribosomes

    C. microtubules

    D. plastids

9. Smooth endoplasmic reticulum is involved in _____.

    A. producing proteins

    B. producing ATP

    C. producing cilia

    D. removing toxins

10. Centrioles are involved in _____.

    A. producing proteins

    B. producing ATP

    C. producing cilia

    D. removing toxins

## Match the Correct Answers

_____ 11. Lysosomes

_____ 12. Grana

_____ 13. Chromoplasts

_____ 14. Contractile vacuole

_____ 15. Cellulose

A. contain the materials necessary for photosynthesis

B. polysaccharide that animals cannot digest

C. release enzymes to digest molecules outside a cell

D. contain red and yellow pigments

E. expels water from protozoa cells

## Match the Correct Answers

_____ 1. Mitosis

_____ 2. Cleavage

_____ 3. Cytokinesis

_____ 4. Zygote

_____ 5. Diploid

A. single cell produced from an egg cell and sperm cell DNA

B. cytoplasm divides

C. division of a zygote

D. cells with 2 of each chromosome

E. cell nucleus divides

_____ 6. Prophase

_____ 7. Anaphase

_____ 8. Metaphase

_____ 9. Cell plate

_____ 10. Telophase

A. chromosomes are pulled toward opposite ends of cell

B. forms in plant cell division

C. chromosomes condense

D. chromosomes at opposite ends of cell

E. chromosomes are pulled to center of cell

_____ 11. Chromosome replicates

_____ 12. Centromere

_____ 13. Interphase

_____ 14. Somatic cells

_____ 15. Virchow's Cell Law

A. cells come from cells

B. forms sister chromatids

C. what happens to a cell between divisions

D. holds sister chromatids together

E. cells not involved in reproduction

## Match the Correct Answers

_____ 1. Microevolution

_____ 2. Habitat

_____ 3. Primary succession

_____ 4. Carrying capacity

_____ 5. Niche

A. all of your needs

B. limit of number of ants an area can support

C. life is established on island formed by a volcano

D. small amount of change within a kind

E. your home


_____ 6. Population

_____ 7. Climax community

_____ 8. Community

_____ 9. Ecology

_____ 10. Secondary succession

A. study of the interrelationships of organisms and their environment

B. organisms in same gene pool

C. plants grow in an area after a fire

D. all living organisms living in the same region

E. process of succession goes undisturbed long enough


_____ 11. Population density

_____ 12. Species

_____ 13. Kinds

_____ 14. Ecosystem

_____ 15. Gene pool

A. biblical created groups

B. all living and non-living things in an area

C. number of same species in an area

D. genetic variation in a population

E. naturally interbreed

## Match the Correct Answers

_____ 1. Deserts

_____ 2. Grasslands

_____ 3. Mountains

_____ 4. Deciduous trees

_____ 5. Coniferous trees

A. exposed to extreme temperatures between seasons

B. are cone-bearing

C. low rainfall totals

D. cause rainstorms to lose rain

E. do not produce cones


_____ 6. Tropical rain forests

_____ 7. Chaparral

_____ 8. Limnology

_____ 9. Oligotrophic

_____ 10. Plankton

A. along west coast of the United States

B. carried by movement of water

C. poor in minerals and other nutrients

D. most productive and diverse biomes on earth

E. study of freshwater biology


_____ 11. Benthos

_____ 12. Nekton

_____ 13. Swamps and marshes

_____ 14. Eutrophic

_____ 15. Deciduous

A. live on or in bottom of aquatic ecosystems

B. some of the richest aquatic habitats for photosynthesis

C. lose leaves in the fall season

D. when water accumulates excessive nutrients

E. can move considerable distances in the direction of their choosing

## Match the Correct Answers

_____ 1. Green                       A. autotrophs

_____ 2. Plants                      B. traps light energy

_____ 3. Animals                   C. heterotrophs

_____ 4. Chlorophyll            D. plants absorb

_____ 5. Blue, violet, red, orange    E. plants reflect


_____ 6. $CO_2$, $H_2O$               A. addition of electrons

_____ 7. $C_6H_{12}O_6$, $O_2$          B. removal of electrons

_____ 8. Oxidation               C. produced in Light Reactions

_____ 9. Reduction              D. used by photosynthesis

_____ 10. ATP                    E. produced by photosynthesis


_____ 11. CAM plants          A. provides hydrogen and electrons for Dark Reactions

_____ 12. $C_4$ plants             B. live under moderate climate conditions

_____ 13. NADPH             C. uses energy from sulfur compounds

_____ 14. $C_3$ plants             D. open stomates only at night

_____ 15. Chemosynthesizers    E. Corn

## Match the Correct Answers

_____ 1. Glycolysis

_____ 2. Glucose

_____ 3. Formation of ATP

_____ 4. Pyruvic acid

_____ 5. Hydrogen atoms

A. formed in glycolysis

B. oxidative phosphorylation

C. anaerobic

D. enters glycolysis

E. removed by NAD


_____ 6. Lactic acid

_____ 7. PGAL

_____ 8. Cristae

_____ 9. Acetyl CoA

_____ 10. $CO_2$

A. inner membrane of mitochondria

B. formed when oxygen is lacking

C. carbon atoms are removed as _____

D. enters Krebs cycle

E. glucose is broken down into _____


_____ 11. Electrons

_____ 12. Hydrogen

_____ 13. Electron Transport System

_____ 14. Oxaloacetic acid

_____ 15. Between 34 and 38

A. where hydrogen atoms are taken

B. oxygen accepts _____

C. number of ATP molecules from each glucose molecule

D. _____ ions pass into the matrix

E. has 4 carbon atoms

**Match the Correct Answers**

_____ 1. Cytosine           A. DNA replication

_____ 2. DNA structure      B. opposite guanine

_____ 3. Thymine            C. sugar, phosphate, base

_____ 4. Nucleotide         D. double helix

_____ 5. Semiconservative   E. opposite adenine


_____ 6. Regulator genes    A. DNA that codes for a protein

_____ 7. Meiosis            B. _____ bonds hold DNA strands together

_____ 8. Genes              C. determines if a gene is turned on or off

_____ 9. Hydrogen           D. base not in RNA

_____ 10. Thymine           E. produces haploid cells

**Circle the Correct Answers**

11. The sugar in RNA is _____.

    A. glucose

    B. ribose

    C. deoxyribose

    D. uracil

12. RNA has the base _____ that is not in DNA.

    A. uracil

    B. thymine

    C. adenine

    D. guanine

13. New DNA has _____.

    A. 2 new strands

    B. 2 old strands

    C. a new strand and an old strand

14. _____ was used to identify new strands of DNA.

    A. $^{15}N$ thymine

    B. $^{12}C$ adenine

    C. uracil

15. $^{14}$N $^{15}$N DNA is heavier than _____.

    A. $^{15}$N $^{15}$N DNA

    B. $^{14}$N $^{15}$N DNA

    C. $^{14}$N $^{14}$N DNA

## THE GENETIC CODE FOR mRNA

| First Base Unit | Second Base Unit | Third Base Unit | | | |
|-----------------|------------------|------|------|------|------|
|  |  | A | G | U | C |
| Adenine | A | lys | lys | asn | asn |
|  | G | arg | arg | ser | ser |
|  | U | ile | met | ile | ile |
|  | C | thr | thr | thr | thr |
| Guanine | A | glu | glu | asp | asp |
|  | G | gly | gly | gly | gly |
|  | U | val | val | val | val |
|  | C | ala | ala | ala | ala |
| Uracil | A | stop | stop | tyr | tyr |
|  | G | stop | trp | cys | cys |
|  | U | leu | leu | phe | phe |
|  | C | ser | ser | ser | ser |
| Cytosine | A | gln | gln | his | his |
|  | G | arg | arg | arg | arg |
|  | U | leu | leu | leu | leu |
|  | C | pro | pro | pro | pro |

## Match the Correct Answers

_____ 1. Codon           A. length of DNA that codes for a protein

_____ 2. Gene           B. refers to the change in allele frequency over generations

_____ 3. RNA           C. length of DNA that codes for an amino acid

_____ 4. Evolution           D. 1 strand

_____ 5. Redundant code           E. more than 1 codon for an amino acid


_____ 6. Alleles           A. rare mutated gene

_____ 7. Genotype           B. variations of the same gene

_____ 8. Phenotype           C. gene that produces a normal protein in a species

_____ 9. Recessive           D. physical expression of genes

_____ 10. Wild type           E. the two copies of each gene for an individual


_____ 11. AUG           A. gly (glycine)

_____ 12. GGG           B. stop

_____ 13. AUC           C. arg (argenine)

_____ 14. UGA           D. start

_____ 15. CGU           E. ile (isoleucine)

## Match the Correct Answers

_____ 1. Nucleus     A. parts of mRNA that are removed

_____ 2. Proteins     B. amino acid chains

_____ 3. Ribosomes    C. parts of mRNA that are spliced together

_____ 4. Introns     D. protects DNA

_____ 5. Exons      E. where amino acids attach together

<br>

_____ 6. Transcribe    A. UCA in mRNA

_____ 7. CTG in DNA   B. can interbreed to form hybrids

_____ 8. AGT in DNA   C. CGU in mRNA

_____ 9. GCA in DNA   D. GAC in mRNA

_____ 10. Species     E. means to copy (rewrite)

<br>

_____ 11. Frameshift    A. different exon combinations

_____ 12. Removal of introns   B. used in Genesis; different from the word species

_____ 13. Alternative splicing   C. occurs if enzymes cut original mRNA in the wrong place

_____ 14. mRNA     D. controlled by enzymes, not random

_____ 15. Kinds      E. a portion of DNA is copied to produce a single strand of _____

## Match the Correct Answers

_____ 1. tRNA

_____ 2. Ribosomes

_____ 3. Translation

_____ 4. Activator enzyme

_____ 5. Polysome

A. process of synthesizing proteins with mRNA and ribosomes

B. partially double stranded

C. attach to mRNA when it leaves the nucleus

D. attaches an amino acid to its appropriate tRNA

E. many ribosomes attached to the same mRNA

_____ 6. P site

_____ 7. A site

_____ 8. tRNA GUA

_____ 9. tRNA CGU

_____ 10. tRNA

A. mRNA CAU

B. mRNA GCA

C. where first amino acid–tRNA combination attaches

D. where first amino acid–tRNA moves to when second one attaches to the ribosome

E. where amino acids are first attached

_____ 11. ATP

_____ 12. N-formyl methionine

_____ 13. Polypeptide

_____ 14. Codon

_____ 15. Irreducible complexity

A. ribosome slides down to the next _____ with each new amino acid–tRNA

B. chain of amino acids

C. first amino acid placed in a protein

D. the interaction between nucleic acids and proteins

E. supplies energy to attach amino acids to each other

## Match the Correct Answers

_____ 1.  Fetus

_____ 2.  Bacteria

_____ 3.  Embryo

_____ 4.  Centromere

_____ 5.  Homeobox genes

A. genes that start the development of arms, legs, eyes, and other important body parts

B. human after 8 weeks of development

C. divide by binary fission

D. divided zygote

E. where spindle fibers attach

_____ 6.  Interphase

_____ 7.  Metaphase

_____ 8.  Prophase

_____ 9.  Telophase

_____ 10. Anaphase

A. nuclear membrane comes apart

B. nuclei reform

C. DNA and organelles replicate

D. chromosomes are pulled toward middle of cell

E. chromosomes are pulled toward sides of cell

_____ 11. Meiosis

_____ 12. Mitosis

_____ 13. Chromosomes

_____ 14. Cytokinesis

_____ 15. Crossing over

A. what sister chromatids are called once they can be seen with a light microscope

B. haploid cells produced

C. division of the cell itself into 2 daughter cells

D. when sister chromatids exchange genes with each other

E. diploid cells produced

## Match the Correct Answers

_____ 1. Punnett Square      A. 100% round peas

_____ 2. rr      B. ¾ round peas

_____ 3. RR x rr      C. wrinkled peas

_____ 4. Rr x Rr      D. used to determine genetic outcomes

_____ 5. Homozygous      E. both alleles are the same


_____ 6. LL x ll      A. very frizzled feathers

_____ 7. Ll x Ll      B. short-haired cats

_____ 8. $F^+F^+$      C. mildly frizzled feathers

_____ 9. $F^+F^f$      D. 25% long-haired cats

_____ 10. $F^fF^f$      E. normal feathers


_____ 11. $F^+F^f$ x $F^+F^f$      A. 50% mildly frizzled feathers

_____ 12. Genotype      B. indicates normal, wild type

_____ 13. Phenotype      C. the genetic pattern

_____ 14. +      D. the physical appearance

_____ 15. Heterozygous      E. the alleles are different

Biology

Genetic Patterns 2

Quiz 20
Lesson 20

Day 125

Total score: _____ of 100

Name

**Match the Correct Answers**

_____ 1. RrYy x RrYy

_____ 2. Law of Independent Assortment

_____ 3. RRYY pea plants

_____ 4. RRyy pea plants

_____ 5. Rr x Rr

A. monohybrid cross

B. all round, green peas

C. all round, yellow peas

D. dihybrid cross

E. allele for a gene does not affect which is inherited for another gene

_____ 6. Y chromosome

_____ 7. Amniocentesis

_____ 8. Synapse

_____ 9. Not significantly different

_____ 10. Tetrad

A. process of drawing out amniotic fluid around a fetus

B. observed and expected difference less than 5%

C. missing some genes that are on the X chromosome

D. formed by the process of synapsis

E. when homologous chromosomes come together

_____ 11. Karyotype

_____ 12. XY

_____ 13. XX

_____ 14. More common in human males

_____ 15. Linked

A. genes on same chromosome

B. human female

C. human male

D. picture of chromosomes

E. hemophilia and color blindness

## Match the Correct Answers

_____ 1. Fixity of species

_____ 2. Point mutations

_____ 3. John Ray

_____ 4. Carolus Linnaeus

_____ 5. Translocations

A. first to use "species" as a biological term

B. species same as kinds

C. parts of chromosomes are moved around

D. change base pairs of DNA

E. placed similar organisms in the same species

_____ 6. Transposons

_____ 7. Inversion

_____ 8. Deletion

_____ 9. Duplication

_____ 10. Polyploidy

A. ABCD → ACD

B. ABCD → ACBD

C. ABCD → ABCCD

D. more than 2 of each chromosome

E. sections of chromosomes that are moved around

_____ 11. Polydactyly

_____ 12. Allopolyploid

_____ 13. Natural selection

_____ 14. Artificial selection

_____ 15. Polyploidy

A. control which plants interbreed

B. plants with larger flowers and fruit; sometimes seedless

C. 6 fingers on each hand

D. plant with all the pairs of chromosomes from both parent plants

E. environment causes some organisms to survive better

## Match the Correct Answers

_____ 1. Genomics

_____ 2. Specific treatments

_____ 3. SNP

_____ 4. Bioinformatics

_____ 5. Genome

A. use of computer software in studying genomics

B. study of base sequences in DNA

C. sequence of genes (or bases) in an organism's chromosomes

D. possible to design by sequencing a person's DNA

E. area in DNA where there is a different base than in most people

_____ 6. Clones

_____ 7. Comparative genomics

_____ 8. HapMap

_____ 9. Aspen trees

_____ 10. Sea anemone

A. study of similarities and differences in genomes

B. invertebrate that naturally forms clones

C. produce many clones asexually

D. section of DNA with SNPs

E. genetically identical organisms

_____ 11. Dolly

_____ 12. Haplotype

_____ 13. Allele

_____ 14. Common functions

_____ 15. Cloning

A. base that differs

B. sequence of SNPs

C. genes in humans and in fruit flies have _____

D. DNA from mature cell placed in egg

E. the first mammal to be cloned

## Match the Correct Answers

| _____ 1. Taxonomy | A. have tubes to carry water and nutrients |
|---|---|
| _____ 2. Vascular plants | B. group within an order |
| _____ 3. Family | C. group within a class |
| _____ 4. Class | D. the study of classification |
| _____ 5. Order | E. group within a phylum |

| _____ 6. Mycorrhizae | A. phylum Sphenophyta |
|---|---|
| _____ 7. Sporophyte | B. phylum Pterophyta |
| _____ 8. Ferns | C. fungi that help move water and nutrients in whisk ferns |
| _____ 9. Horsetails | D. non-vascular plants |
| _____ 10. Bryophytes | E. Bryophyte diploid structures |

| _____ 11. Club mosses | A. seedless vascular plants |
|---|---|
| _____ 12. Whisk ferns | B. divided into the classes Monocotyledonae and Dicotyledonae |
| _____ 13. Phylum Magnoliophyta | C. monocots |
| _____ 14. Gymnosperms | D. phylum Psilophyta |
| _____ 15. Grasses | E. another name for conifers |

## Match the Correct Answers

_____ 1. Colony

_____ 2. Multicellular animals

_____ 3. Comparative anatomy

_____ 4. Cladistics

_____ 5. Baraminology

A. have different cells carrying out different roles

B. only selected characteristics are used for classification

C. the study of grouping organisms in their kind of creation

D. cells can handle all necessary functions when not assembled together

E. traditionally used almost exclusively in classifying organisms

_____ 6. Species

_____ 7. Taxon

_____ 8. Sponges

_____ 9. Flatworms

_____ 10. Jellyfish

A. can interbreed

B. phylum Porifera

C. recognized classification group

D. phylum Cnidaria

E. phylum Platyhelminthes

_____ 11. Roundworms

_____ 12. Earthworms

_____ 13. Spiders

_____ 14. Polyp

_____ 15. Starfish

A. upright body form

B. phylum Echinodermata

C. phylum Annelida

D. class Arachnida

E. phylum Nematoda

## Match the Correct Answers

_____ 1. Notochord

_____ 2. Phylum Chordata

_____ 3. Gills

_____ 4. Hagfish

_____ 5. Ostracoderms

A. rod of cartilage that vertebrae develop around in most vertebrates

B. extinct jawless fish

C. class Agnatha

D. pharyngeal pouches in fish and tadpoles

E. vertebrates


_____ 6. Oviparous

_____ 7. Bony fish

_____ 8. Viviparous

_____ 9. Sharks

_____ 10. Dinosaurs

A. fish embryos that develop outside mother's body

B. fish embryos that develop within mother's body (with nourishment from her body)

C. class Reptilia

D. class Osteichthyes

E. class Chondrichthyes


_____ 11. Fish

_____ 12. subclass Prototheria

_____ 13. Salamanders

_____ 14. Birds

_____ 15. Mammals

A. order Urodela

B. class Aves

C. 1 atrium and 1 ventricle

D. egg-laying mammals

E. class Mammalia

## Match the Correct Answers

_____ 1. William Smith

_____ 2. Lucretius

_____ 3. Lamarck

_____ 4. James Hutton

_____ 5. Charles Darwin

A. pangenesis

B. small changes over millions of years make large changes

C. uniformitarianism

D. rock layers dated by fossils

E. earth itself yielded plants and animals

_____ 6. Charles Lyell

_____ 7. Microevolution

_____ 8. Agnostic Period

_____ 9. Punctuated equilibrium

_____ 10. BioLogos

A. evolution occurred in rapid events

B. first major textbook in geology

C. if evolution were not true, people didn't know where life came from

D. supports theistic evolution

E. small changes in existing DNA

_____ 11. Gap Theory

_____ 12. Day Age Theory

_____ 13. Metamorphic rock

_____ 14. Neo-Darwinism

_____ 15. Cambrian Explosion

A. vast ages of time took place between Genesis 1:1 and 1:2

B. 90% of fossils showing evolution are missing

C. survival of those who can reproduce offspring that survive

D. days of creation overlap

E. formed by massive pressures and high temperatures

## Match the Correct Answers

_____ 1. Convergent evolution     A. used to indicate relatedness and evolutionary time

_____ 2. Niches     B. similar structures evolved more than once in different organisms

_____ 3. Mammals     C. created on fifth day of creation

_____ 4. Birds     D. living creatures were created to fill all available _____

_____ 5. Molecular clock     E. created on sixth day of creation

 

_____ 6. mtDNA     A. human tailbones were thought to be _____

_____ 7. Radioisotope decay     B. compared between organisms to determine how long it took them to evolve

_____ 8. Vestigial organs     C. usually appear in the same order

_____ 9. Fossils     D. ages assigned to the geological column

_____ 10. Mount St. Helens     E. helped replace opinion with observable, measurable data

 

_____ 11. Data     A. convinced most that earth was old

_____ 12. Data application     B. estimate of the life forms created that are still around

_____ 13. Fossils in order     C. same for creationists and evolutionists

_____ 14. 10%     D. usually do not interbreed with each other

_____ 15. Different species     E. different for creationists and evolutionists

## Match the Correct Answers

Questions 1–5 refer to human classification.

_____ 1. Chordata               A. class

_____ 2. Eukarya                B. phylum

_____ 3. Hominidae              C. order

_____ 4. Primate               D. family

_____ 5. Mammalia              E. domain


_____ 6. Neanderthal            A. southern ape

_____ 7. Peking man             B. posture was based on bones of diseased and elderly forms

_____ 8. Cro-Magnon             C. hoax

_____ 9. *Australopithecus*       D. skull volumes less than average of humans today

_____ 10. Piltdown man          E. conlcuded to be *Homo sapiens*


_____ 11. Mitochondria          A. nicknamed Lucy

_____ 12. Conserved traits      B. inherited from mother through egg

_____ 13. *Australopithecus afarensis*   C. evolutionary idea of an early primate

_____ 14. Prosimian             D. common DNA among most life forms

_____ 15. Fossil data           E. does not, and will not, contradict the Bible

# Exams

## Match the Correct Answers

| | | |
|---|---|---|
| _____ 1. Glucose | A. metal and non-metal |
| _____ 2. Metal | B. 6 carbon atoms |
| _____ 3. Non-metal | C. gains electrons |
| _____ 4. Covalent bond | D. loses electrons |
| _____ 5. Ionic bond | E. non-metal and non-metal |

| | | |
|---|---|---|
| _____ 6. Isotopes | A. have 8 electrons in outer shell |
| _____ 7. Atomic number | B. has + and – ends |
| _____ 8. More stable atoms | C. same number of protons, different number of neutrons |
| _____ 9. Water | D. allow water to dissolve polar compounds |
| _____ 10. Hydrogen bonds | E. the number of an atom's protons |

| | | |
|---|---|---|
| _____ 11. pH 3 | A. base |
| _____ 12. pH 9 | B. less dense than water |
| _____ 13. $^-$OH | C. what Schrodinger called the energy levels of electrons |
| _____ 14. Ice | D. more base than acid |
| _____ 15. Orbitals | E. more acid than base |

| | | |
|---|---|---|
| _____ 16. Sucrose | A. animal starch |
| _____ 17. Glycogen | B. no double bonds |
| _____ 18. Triglycerides | C. polysaccharide |
| _____ 19. Cellulose | D. glucose and fructose bonded together |
| _____ 20. Saturated fatty acid | E. animal fat |

| | | |
|---|---|---|
| _____ 21. Starch | A. an energy storage molecule |
| _____ 22. Adipose tissue | B. an acid and the carboxylic end of an amino acid |
| _____ 23. Protein | C. in RNA but not in DNA |
| _____ 24. –COOH | D. excess energy stored around the body |
| _____ 25. Uracil | E. chain of amino acids |

_____ 26. Ribonucleic acid

_____ 27. Thymine

_____ 28. Sugar and phosphate

_____ 29. ATP

_____ 30. Sequence of amino acids

A. backbone of nucleic acid

B. primary structure of protein

C. RNA

D. stores energy in cells

E. in DNA but not in RNA

## Match the Correct Answers

_____ 1. Virchow

_____ 2. Robert Hooke

_____ 3. Dutrochet

_____ 4. Schleiden and Schwann

_____ 5. Hans Spemann

A. saw that plant tissues had cells

B. found that the nucleus determines what organism looks like

C. named after holes in cork

D. proposed the Cell Law

E. demonstrated that cells come from other cells

_____ 6. Eukaryotes

_____ 7. Prokaryotes

_____ 8. Ultracentrifuge

_____ 9. Millimeter

_____ 10. Micrometer

A. $10^{-6}$ meter

B. $10^{-3}$ meter

C. do not have nuclear membranes

D. have nuclear membranes

E. pulls heavier parts of cells away from others

_____ 11. Confocal microscope

_____ 12. Cell membranes

_____ 13. Selectively permeable

_____ 14. Chromatin

_____ 15. Diffusion

A. 2 layers of phospholipids

B. uses laser beams

C. move from region of more concentration to less

D. membranes control what passes into or out of cells

E. DNA and its associated proteins together

_____ 16. Nucleolus

_____ 17. Solute

_____ 18. Isotonic

_____ 19. Hypotonic

_____ 20. Membrane pumps

A. physiological saline

B. distilled water

C. contains RNA and protein

D. what NaCl is called when dissolved into water

E. some proteins in the phospholipid layers of a cell membrane

_____ 21. Swell up and burst

_____ 22. Shrivel up

_____ 23. Chloroplasts

_____ 24. Cilia

_____ 25. Centrioles

A. barrel-shaped organelles in animal cells

B. cells in 10% NaCl

C. cells in distilled water

D. small hair-like structures

E. where photosynthesis occurs in plants

_____ 26. Ribosomes

_____ 27. Mitochondria

_____ 28. Golgi bodies

_____ 29. Smooth endoplasmic reticulum

_____ 30. Lysosomes

A. where proteins are formed

B. extension of the membranes of the endoplasmic reticulum

C. release enzymes to digest molecules outside a cell

D. where ATP is formed

E. removes toxins

## Match the Correct Answers

_____ 1. Cell plate

_____ 2. Zygote

_____ 3. Mitosis

_____ 4. Interphase

_____ 5. Cytokinesis

A. cytoplasm divides

B. forms in plant cell division

C. what happens to a cell between divisions

D. single cell produced from an egg cell and sperm cell DNA

E. cell nucleus divides

_____ 6. Metaphase

_____ 7. Prophase

_____ 8. Telophase

_____ 9. Centromere

_____ 10. Climax community

A. chromosomes condense

B. process of succession goes undisturbed long enough

C. holds sister chromatids together

D. chromosomes are pulled to center of cell

E. chromosomes at opposite ends of cell

_____ 11. Niche

_____ 12. Population

_____ 13. Secondary succession

_____ 14. Population density

_____ 15. Carrying capacity

A. plants grow in an area after a fire

B. number of same species in an area

C. all your needs for survival

D. limit of number of fleas an area can support

E. organisms in same gene pool

_____ 16. Deciduous trees

_____ 17. Coniferous trees

_____ 18. Eutrophic

_____ 19. Benthos

_____ 20. Oligotrophic

A. are cone-bearing

B. lose leaves in fall

C. poor in minerals and other nutrients

D. when water accumulates excessive nutrients

E. live on or in bottom of aquatic ecosystems

_____ 21. Mountains

_____ 22. Grasslands

_____ 23. Chlorophyll

_____ 24. Plants

_____ 25. Oxidation

A. traps light energy

B. removal of electrons

C. exposed to extreme temperatures between seasons

D. autotrophs

E. cause rainstorms to lose rain

_____ 26. Green

_____ 27. $C_6H_{12}O_6$ and $O_2$

_____ 28. NADPH

_____ 29. CAM plants

_____ 30. $C_3$ plants

A. open stomates only at night

B. plants reflect

C. produced by photosynthesis

D. live under moderate climate conditions

E. provides hydrogen and electrons for Dark Reactions

## THE GENETIC CODE FOR mRNA

| First Base Unit | Second Base Unit | Third Base Unit | | | |
|---|---|---|---|---|---|
| | | **A** | **G** | **U** | **C** |
| Adenine | A | lys | lys | asn | asn |
| | G | arg | arg | ser | ser |
| | U | ile | met | ile | ile |
| | C | thr | thr | thr | thr |
| Guanine | A | glu | glu | asp | asp |
| | G | gly | gly | gly | gly |
| | U | val | val | val | val |
| | C | ala | ala | ala | ala |
| Uracil | A | stop | stop | tyr | tyr |
| | G | stop | trp | cys | cys |
| | U | leu | leu | phe | phe |
| | C | ser | ser | ser | ser |
| Cytosine | A | gln | gln | his | his |
| | G | arg | arg | arg | arg |
| | U | leu | leu | leu | leu |
| | C | pro | pro | pro | pro |

## Match the Correct Answers

_____ 1. Glucose

_____ 2. Pyruvic acid

_____ 3. Acetyl CoA

_____ 4. Glycolysis

_____ 5. Lactic acid

A. enters Krebs cycle

B. enters glycolysis

C. formed in glycolysis

D. formed when oxygen is lacking

E. anaerobic


_____ 6. Electrons

_____ 7. Formation of ATP

_____ 8. Nucleotide

_____ 9. $CO_2$

_____ 10. Semiconservative

A. carbon atoms removed as _____

B. DNA replication

C. sugar, phosphate, base

D. oxidative phosphorylation

E. accepted by oxygen atoms

_____ 11. New DNA              A. length of DNA that codes for a protein

_____ 12. Uracil                  B. opposite guanine

_____ 13. Regulator genes     C. base in RNA but not in DNA

_____ 14. Cytosine             D. has an old strand and a new strand

_____ 15. Gene                   E. determines if a gene is turned on or off


_____ 16. Evolution           A. length of DNA that codes for an amino acid

_____ 17. Redundant code     B. gly (glycine)

_____ 18. GGA                 C. refers to the change in allele frequency over generations

_____ 19. AUC                 D. more than 1 codon for an amino acid

_____ 20. Codon               E. ile (isoleucine)


_____ 21. AUG                 A. his (histidine)

_____ 22. Ribosomes         B. parts of mRNA that are removed

_____ 23. Introns             C. start

_____ 24. CAC                 D. parts of mRNA that are spliced together

_____ 25. Exons                E. where amino acids attach together


_____ 26. AAG in DNA       A. ACU in mRNA

_____ 27. Frameshift         B. different exon combinations

_____ 28. Alternative splicing   C. occurs if enzymes cut original mRNA in the wrong place

_____ 29. TGA in DNA       D. UUC in mRNA

_____ 30. mRNA               E. a portion of DNA is copied to produce a single strand of _____

## Match the Correct Answers

_____ 1. ATP

A. where first amino acid–tRNA combination attaches

_____ 2. Translation

B. where first amino acid–tRNA moves to when second one attaches to the ribosome

_____ 3. A site

C. process of synthesizing proteins with mRNA and ribosomes

_____ 4. P site

D. supplies energy to attach amino acids to each other

_____ 5. Activator enzyme

E. attaches an amino acid to its appropriate tRNA

_____ 6. tRNA CCU

A. divide by binary fission

_____ 7. tRNA GAC

B. mRNA CUG

_____ 8. N-formyl methionine

C. mRNA GGA

_____ 9. tRNA

D. first amino acid placed in a protein

_____ 10. Bacteria

E. where amino acids are first attached

_____ 11. Meiosis

A. diploid cells produced

_____ 12. Crossing over

B. divided zygote

_____ 13. Mitosis

C. where spindle fibers attach

_____ 14. Centromeres

D. haploid cells produced

_____ 15. Embryo

E. when sister chromatids exchange genes with each other

_____ 16. Cytokinesis

A. chromosomes are pulled toward sides of cell

_____ 17. Anaphase

B. DNA and organelles replicate

_____ 18. Interphase

C. division of the cell itself into 2 daughter cells

_____ 19. Homeobox genes

D. nuclei reform

_____ 20. Telophase

E. genes that start the development of arms, legs, eyes, and other important body parts

_____ 21. Homozygous

A. both alleles are the same

_____ 22. RR x rr

B. 100% short-haired cats

_____ 23. +

C. indicates normal, wild type

_____ 24. Ll x Ll

D. 100% round peas

_____ 25. LL x ll

E. 25% long-haired cats

_____ 26. Heterozygous

_____ 27. $F^+F^f$

_____ 28. Genotype

_____ 29. $F^+F^f$ x $F^+F^f$

_____ 30. $F^+F^+$

A. mildly frizzled feathers

B. the genetic pattern

C. the alleles are different

D. 50% mildly frizzled feathers

E. normal feathers

## Match the Correct Answers

_____ 1. Tetrad

_____ 2. Karyotype

_____ 3. Law of Independent Assortment

_____ 4. Synapse

_____ 5. Linked

A. genes on same chromosome

B. formed by the process of synapsis

C. when homologous chromosomes come together

D. picture of chromosomes

E. allele of a gene does not affect which allele is inherited for another gene

_____ 6. XX

_____ 7. XY

_____ 8. Amniocentesis

_____ 9. More common in human males

_____ 10. Polyploidy

A. human male

B. human female

C. hemophilia and color blindness

D. more than 2 of each chromosome

E. process of drawing out amniotic fluid around a fetus

_____ 11. RRyy pea plants

_____ 12. Rr x Rr

_____ 13. Translocation

_____ 14. Carolus Linnaeus

_____ 15. John Ray

A. first to use "species" as a biological term

B. placed similar organisms in the same species

C. all round, green peas

D. parts of chromosomes are moved around

E. monohybrid cross

_____ 16. Artificial selection

_____ 17. Transposons

_____ 18. Inversion

_____ 19. Duplication

_____ 20. Polydactyly

A. sections of chromosomes that are moved around

B. control which plants interbreed

C. ABCD → ACBD

D. 6 fingers on each hand

E. ABCD → ABCDD

_____ 21. Genomics

_____ 22. Genome

_____ 23. Haplotype

_____ 24. Bioinformatics

_____ 25. Clones

A. sequence of SNPs

B. use of computer software in studying genomics

C. sequence of genes (or bases) in an organism's chromosomes

D. study of base sequences in DNA

E. genetically identical organisms

_____ 26. HapMap

_____ 27. Aspen trees

_____ 28. SNP

_____ 29. Cloning

_____ 30. Common functions

A. produce many clones asexually

B. DNA from mature cell placed in egg

C. area in DNA where there is a different base than in most people

D. section of DNA with SNPs

E. genes in humans and fruit flies have _____

## Match the Correct Answers

_____ 1. Class               A. species name

_____ 2. Family           B. group within a phylum

_____ 3. Order              C. group within a class

_____ 4. *Homo*             D. group within an order

_____ 5. *sapiens*           E. genus name

_____ 6. Monocots       A. another name for conifers

_____ 7. Gymnosperms   B. seedless vascular plants

_____ 8. Club mosses      C. non-vascular plants

_____ 9. Ferns              D. grasses

_____ 10. Bryophytes      E. phylum Pterophyta

_____ 11. Cladistics        A. class Arachnida

_____ 12. Baraminology    B. only selected characteristics are used for classification

_____ 13. Colony           C. recognized classification group

_____ 14. Spiders          D. cells can live independently

_____ 15. Taxon            E. the study of grouping organisms in their kind of creation

_____ 16. Roundworms    A. phylum Annelida

_____ 17. Earthworms     B. phylum Porifera

_____ 18. Flatworms       C. phylum Nematoda

_____ 19. Sponges         D. phylum Cnidaria

_____ 20. Jellyfish          E. phylum Platyhelminthes

_____ 21. Hagfish          A. egg-laying mammals

_____ 22. Oviparous       B. pharyngeal pouches in fish and tadpoles

_____ 23. Viviparous       C. class Agnatha

_____ 24. Gills              D. fish embryos that develop within the mother's body (with nourishment from her body)

_____ 25. subclass Prototheria    E. fish embryos that develop outside the mother's body

_____ 26. Bony fish          A.  order Urodela

_____ 27. Sharks             B.  class Osteichthyes

_____ 28. Salamanders        C.  class Chondrichthyes

_____ 29. Mammals            D.  class Aves

_____ 30. Birds              E.  class Mammalia

## Match the Correct Answers

_____ 1. Day Age Theory      A. survival of those who can reproduce offspring that survive

_____ 2. Gap Theory      B. vast ages of time took place between Genesis 1:1 and 1:2

_____ 3. James Hutton      C. days of creation overlap

_____ 4. Neo-Darwinism      D. first major geology textbook

_____ 5. Charles Lyell      E. uniformitarianism

_____ 6. Punctuated equilibrium      A. 90% of fossils showing evolution are missing

_____ 7. Cambrian Explosion      B. pangenesis

_____ 8. Charles Darwin      C. evolution occurred in rapid events

_____ 9. Microevolution      D. small changes in existing DNA

_____ 10. Lamarck      E. small changes over millions of years make large changes

_____ 11. Molecular clock      A. compared between organisms to determine how long it took them to evolve

_____ 12. Mammals      B. created on the sixth day of creation

_____ 13. Birds      C. created on the fifth day of creation

_____ 14. Vestigial organs      D. used to indicate relatedness and evolutionary time

_____ 15. mtDNA      E. human tailbones were thought to be _____

_____ 16. Mount St. Helens      A. ages assigned to the geological column

_____ 17. Niches      B. helped replace opinion with observable, measurable data

_____ 18. Radioisotope decay      C. formed by massive pressures and high temperatures

_____ 19. Metamorphic rock      D. living creatures were created to fill all available _____

_____ 20. Fossils in order      E. convinced most that earth was old

_____ 21. Hominidae      A. order

_____ 22. Primate      B. hoax

_____ 23. Mammalia      C. family

_____ 24. Neanderthal      D. posture was based on bones of diseased and elderly forms

_____ 25. Piltdown man      E. class

_____ 26. *Australopithecus afarensis*    A.  does not, and will not, contradict the Bible

_____ 27. Fossil data    B.  skull volumes less than average of humans today

_____ 28. Peking man    C.  evolutionary idea of an early primate

_____ 29. Cro-Magnon    D.  nicknamed Lucy

_____ 30. Prosimian    E.  concluded to be *Homo sapiens*

# Answers to Worksheets,
## Laboratory Reports,
## Quizzes
## and
## Exams

# Biology → Worksheet and Lab Report Answer Keys

**Lesson 1**
**Worksheet 1**

1. Element
2. Protons, neutrons, electrons (in any order)
3. Protons
4. Element
5. Molecule
6. 6
7. 6
8. 12
9. 6
10. Protons, neutrons (in any order)
11. A two-dimensional picture of the arrangement of the atoms in a molecule.
12. $12 H_2O + 6 CO_2 \rightarrow C_6H_{12}O_6 + 6 O_2 + 6 H_2O$
13. B
14. C
15. A

**Lesson 1**
**Worksheet 2**

1. +1
2. Neutral
3. –1
4. Protons, neutrons (any order)
5. Protons, neutrons (any order)
6. Protons, neutrons (in this order)
7. Shells
8. 2
9. Yes. They both have 6 protons and they have different numbers of neutrons.
10. No. They have different numbers of protons.
11. B
12. A
13. E
14. C
15. D

**Lesson 1**
**Lab Report**

The student is to go to a place outdoors and describe the observed life forms. The report is to consist of complete sentences and clear descriptions. Award a possible 20 points for the overall descriptions:

- 8 points for identifying and describing a keystone species (one that if removed would have a large impact on the others)
- 6 points for identifying and describing consumers (feeders)
- 6 points for identifying and describing decomposers. The impacts of bacterial decomposition will be seen rather than the bacteria themselves. Mold also counts as a decomposer.

**Lesson 2**
**Worksheet 1**

1. Negative, positive (in this order)
2. Attracted
3. Hydrogen
4. Non-polar
5. Hydrophilic, hydrophobic
6. the amount of heat energy necessary to raise 1 gram of water one degree Celsius
7. C
8. A
9. D
10. B

**Lesson 2**
**Worksheet 2**

1. 10 million, $H^+$ and $^-OH$
2. $H^+$
3. $^-OH$
4. Neutral
5. 100
6. 10
7. D
8. A

9. B

10. C

## Lesson 2
## Lab Report

When grading the lab report, award

- 10 possible points for following directions

- 10 points for answering the questions for the pH procedures and hydrogen bonding procedures

The answers are the student's observations. 20 points possible overall.

## Lesson 3
## Worksheet 1

1. Monomers

2. Polymers

3. Monosaccharide

4. Disaccharide

5. Polysaccharides

6. Fruit

7. Isomer, atom, arrangement

8. Glucose, fructose (either order)

9. 6-carbon, monosaccharide

10. 12-carbon, disaccharide

11. Glucose, glucose

12. Glucose, galactose (either order)

13. Milk

14. Amylose, amylopectin (either order)

15. C

16. A

17. D

18. B

19. $C_6H_{12}O_6$

## Lesson 3
## Worksheet 2

1. Chitin

2. Fat, oil (either order)

3. Hydrophobic

4. Non-polar

5. Triglycerides

6. Glycerol

7. Fatty acids

8. D

9. A

10. C

11. B

12. Higher melting point and are solid at room temperatures

13. The excess is stored first in the form of glycogen in the liver and then in adipose tissue around the body

## Lesson 3
## Lab Report

The report for steps 1–8 describes the results for test tubes 1–4. The answers to the questions are the student's observations. Award

- 15 possible points for steps 1–8

- 5 possible points for the answers (student observations) for step 9. Iodine turns food dark blue or black when starch is present

20 points possible overall.

## Lesson 4
## Worksheet 1

1. Muscle tissue, cell membranes, and enzymes (any order)

2. Amino acids

3. Amino, carboxyl (in this order)

4. 20

5. Atoms

6. Glycine, H (hydrogen)

7. B

8. A

9. C

10. Long chain of amino acids

11. Coiling of a polypeptide chain into a spiral helix

12. The coiling of a polypeptide chain into a spiral helix

13. When polypeptide chains lie side by side forming a sheet-like structure

**Lesson 4**
**Worksheet 2**

1. Enzymes

2. Substrate

3. Blueprints, managers

4. Nucleotides

5. 5-carbon sugar phosphate, organic base

6. Deoxyribonucleic acid

7. Ribonucleic acid

8. DNA, RNA (any order)

9. Sugar, phosphate

10. 2

11. 1

12. Adenine, guanine, cytosine (any order)

13. Thymine

14. Uracil

15. C

16. D

17. A

18. B

19. Adenosine triphosphate is a nucleotide used to store energy in cells.

**Lesson 4**
**Lab Report**

When grading this lab exercise, give

- 10 points if the instructions were followed

- 10 points if the procedures were done carefully and neatly

20 points possible overall. There is not a report due for this lab. Do not take off points if the DNA extraction did not turn out as well as expected.

**WARNING:** Keep the ethanol away from an open flame. It is important not to place any of the ethanol in the mouth because it may contain a small amount of methanol which is poisonous. It will not be a problem if it touches the skin.

**Lesson 5**
**Worksheet 1**

1. C

2. A

3. D

4. B

5. All living things are composed of cells.

6. It is an observation for which no exceptions were found.

7. New cells were always derived from existing cells of the same kind of organism.

8. Eukaryotes

9. Prokaryotes

10. Bacteria

11. Organelles

12. Organism

13. Organs

14. Tissues

**Lesson 5**
**Worksheet 2**

1. D

2. C

3. A

4. E

5. B

6. If 2 dots were separated by a distance of 0.17 um, they could be seen as 2 separate dots.

7. Laser beams

8. 0.4 nm

9. Surface

10. Through

11. Magnified

12. Cell fractionation

13. Ultracentrifuge

**Lesson 5**
**Lab Report**

When grading this lab exercise, give

- 10 possible points for following instructions and demonstrating good use of the microscope

- 10 possible points for answering all of the questions in the written lab report

20 points possible overall. The answers to the questions are the student's observations. When

looking at the threads through the microscope, the student should be able see which threads are on top of each other by focusing up and down with the fine focusing knob.

## Lesson 6
### Worksheet 1

1. B
2. C
3. A
4.

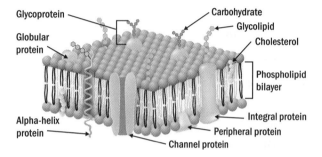

5. Protein and phospholipid molecules are able to move sideways in the membrane.

## Lesson 6
### Worksheet 2

1. Change shape
2. Nucleus
3. Genes
4. Chromosomes
5. Chromatin
6. Nucleolus
7. Cytoplasm
8. Nucleus
9. *Acetabularia*
10. combined some of the characteristics of both species
11. B
12. A
13. C

## Lesson 6
### Lab Report

When grading this lab exercise, grant
• 5 possible points for following directions

• 5 points each for student showing cheek cells, *Spirogyra*, and *Paramecium* under the microscope and describing them

20 points possible overall.

## Lesson 7
### Worksheet 1

1. More, less
2. Solvent, solute
3. Osmosis
4. B
5. D
6. A
7. C
8. The additional water molecules in the confined space of a cell produce pressure against the inside walls of the cell called turgor pressure.

## Lesson 7
### Worksheet 2

1. The cell is active in moving molecules from where they are less concentrated to where they are more concentrated.
2. Membrane
3. Sodium
4. living
5. $Na^+$ (sodium)
6. $Na^+$ (sodium), $K^+$ (potassium)
7. B
8. D
9. C
10. A

## Lesson 7
### Lab Report

Award 20 points possible overall.

• 5 possible points for the answers (student observations) for steps 8 and 9

• 5 possible points each in step 10 for each of parts A, B, and C

In part A of step 10, potato and apple pieces are placed in distilled water. Distilled water is 100%

water and the cells of the potato and apple have 99% water. More water should go into the potato and apple cells (100% into 99%), causing them to swell.

When the potato and apple pieces are placed in 1% NaCl (99% water), the same amount of water should go into the potato and apple cells as go out because the water is the same concentration on both sides of their cell membranes. The potato and apple should not shrink or swell.

When the potato and apple pieces are placed in 10% NaCl (90% water) more water will come out of the potato and apple cells than goes in (99% into 90%). The should cause the potato and apple pieces to shrivel up and shrink.

Award full credit if the student reports what he/she actually observed when directions are carefully followed.

## Lesson 8
### Worksheet 1

1. Centrioles, animal
2. 2
3. Opposite
4. Microtubules
5. Flagella
6. Cilia
7. Mitochondria
8. Cristae
9. Bacteria
10. Endosymbiotic
11. Replicate, abundant
12. Ribosomes
13.

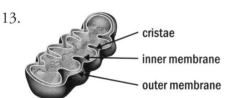

14. B
15. D
16. E
17. A
18. C

## Lesson 8
### Worksheet 2

1. Lysomes release their enzymes to digest molecules outside of a cell; like fungus when it digests whatever it is growing on.
2. D
3. A
4. B
5. C
6. Leucoplasts
7. Chromoplasts
8. Cell walls
9. Cellulose
10. Central vacuole
11. Contractile vacuoles

## Lesson 8
### Lab Report

When grading this lab, grant a possible

• 5 points for each of the four prepared slides as per the directions (20 points possible overall)

Human red blood cells do not have nuclei, which frees up more space to transport oxygen. Frog red blood cells do have nuclei because they do not need to carry as much oxygen because frogs get a lot of oxygen through their skin. If they get too much oxygen, it will start to destroy the frog's cells.

## Lesson 9
### Worksheet 1

1. cells always come from existing cells
2. Zygote
3. Cleavage, divided
4. DNA, copied
5. Unravel, complimentary
6. Old, new (either order)
7. Cell cycle
8. Interphase
9. 3, $G_1$, S, $G_2$
10. Grows
11. DNA, organelles
12. Prepares, divide

13. Mitosis

14. Cytokinesis

15. B

16. A

17. D

18. C

## Lesson 9
## Worksheet 2

1. B

2. D

3. C

4. A

5. Spindle fibers

6. 4

7. Metaphase, center

8. Anaphase, opposite

9. Telophase, opposite, nuclei

10. 2, 2

11. Division ring, halves

12. Cell plate

13. Number, chromosomes

14. 4n, 2, 2n

15. "Behold, I was brought forth in iniquity, and in sin my mother conceived me."

## Lesson 9
## Lab Report

Your student is to find cells in each stage of mitosis (prophase, metaphase, anaphase, and telophase) on the prepared slide of the onion root tip mitosis.

- Award 4 possible points (total 16 points) for finding cells on the slide at each stage of mitosis, and the student's oral description of each stage.

- Award 4 possible points for showing you as many stages of mitosis that he/she finds on the prepared *Ascaris* roundworm slide.

20 points possible overall. This is an oral report. You can find details of each stage in the text in chapter 9.

## Lesson 10
## Worksheet 1

1. groups of similar organisms that naturally interbreed in their natural environment and have offspring that can interbreed and have offspring.

2. Ecology

3. Kinds

4. Population

5. B

6. E

7. A

8. C

9. D

## Lesson 10
## Worksheet 2

1. Habitat

2. Niche

3. Population density

4. Carrying capacity

5. C

6. A

7. B

8. If the process of succession is undisturbed long enough

9. Where deciduous (lose their leaves in the winter) hardwood trees flourish where there had been barren rock or open water

## Lesson 10
## Lab Report

Base the grade of the lab report out of 20 possible points on how well the observations are described using complete sentences. Each step asks for different observations. The answers will depend upon where the observations are made, as they could vary a lot from one place to another.

This is as much a writing assignment as an observation assignment. This is based upon a major goal of biology to understand living organisms in their natural environment.

Step 7 is optional. It involves preparing a fossil that will be dug up and evaluated later in laboratory exercise 26. If this is completed, it will be worth an additional 20 extra credit points when lab exercise 26 is done. These 20 points will raise the total possible points at grading time by 20 points.

**Lesson 11**
**Worksheet 1**

1. Biomes
2. Temperatures
3. Wind, water
4. Bison, cattle (in this order)
5. Desert
6. Mountains
7. Dew
8. The kangaroo rat and a few other desert mammals can live without drinking any water. They are equipped to conserve and use water that they produce during respiration called metabolic water.
9. C
10. D
11. A
12. B

**Lesson 11**
**Worksheet 2**

1. When the soil is cleared for agricultural purposes, heavy rainfall removes minerals and other nutrients from the soil. Tropical forests do not have the sustained fertility found in grasslands and deciduous forests. It is important to protect these forests because many species exist only there, including plants that produce many of our medicines.
2. Productive, diverse (any order)
3. Decomposition
4. Fire
5. Chaparral
6. Xerophytic
7. 10 feet (3 m)
8. Fire
9. Mediterranean
10. Limnology

11. Benthos
12. Nekton
13. Oligotrophic
14. Eutrophic
15. Swamps, marshes
16. D
17. A
18. B
19. C

**Lesson 11**
**Lab Report**

The lab report is worth a total of 20 possible points. Grade the report based upon how well the student reports observations and researches local information. Complete sentences are to be used throughout the report. This is also as much an exercise in writing as it is in gaining information. You can give full credit if everything is completed well.

**Lesson 12**
**Worksheet 1**

1. Plants are producers because they produce food molecules by capturing energy from the sun by photosynthesis. Plants are also called autotrophs, meaning to feed on one's self because they make their own food.
2. D
3. B
4. E
5. A
6. C
7. Chlorophyll
8. Blue, violet, red, orange, green (in this order)
9. Irreducible complexity
10. Chemosynthesizers
11. Second Law of Thermodynamics
12. Oxidation, reduction
13. Oxidation
14. Reduction
15. Reduction
16. Oxidization

## Lesson 12
## Worksheet 2

1. ATP, NADPH, $O_2$ (in any order)

2. ATP

3. NADPH

4. $C_3$

5. $C_3$

6.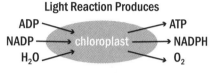

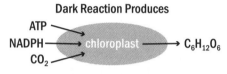

7. D

8. B

9. A

10. C

## Lesson 12
## Lab Report

The lab report is to be graded based upon

• 10 possible points for following directions and correctly setting up the lab

• 10 possible points for describing observations and answering questions with complete sentences

20 points possible overall.

## Lesson 13
## Worksheet 1

1. $C_6H_{12}O_6 + 6 O_2 \rightarrow 6 CO_2 + 6 H_2O + Heat + ATP$

2. Mitochondria

3. Glycolysis, Citric Acid Cycle, Electron Transport System (in this order)

4. Glycolysis

5. Anaerobic

6. Citric Acid, Electron Transport, aerobic (in this order)

7. Oxidative phosphorylation

8. H (hydrogen)

9. Oxygen debt, lactic acid (in this order)

10. C

11. A

12. D

13. B

## Lesson 13
## Worksheet 2

1. $CO_2$

2. Acetyl, oxaloacetic acid, citric acid (in this order)

3. 6, 5, 4

4. Krebs, H (hydrogen) (in this order)

5. NADH, $FADH_2$, $H^+$

6. ATP synthase, chemiosmosis

7. C

8. A

9. B

10.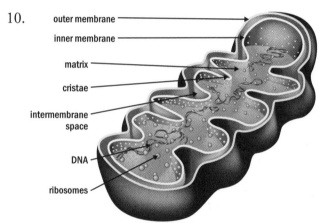

## Lesson 13
## Lab Report

The lab report is to be graded based upon

• 10 possible points for following directions and correctly setting up the lab

• 10 points for answering questions and describing observations with complete sentences

20 points possible overall.

## Lesson 14
## Worksheet 1

1. In the 1800s, objects were observed in the nuclei of cells that absorbed dark stains. Chromo means

color and soma means body — so they were called colored bodies or chromosomes.

2. C

3. A

4. D

5. B

6. Nucleotide

7. Cytosine

8. Guanine

9. Thymine

10. Adenine

11. Hydrogen

12. Semiconservative

13. New, old (any order)

**Lesson 14**
**Worksheet 2**

1. Thymine

2. $^{14}N$, $^{15}N$ (in this order)

3. Quicker, farther (either order)

4. Meiosis, reduction

5. Haploid

6. D

7. C

8. B

9. A

10. The genes that code for the enzymes are usually in the same order in a chromosome.

**Lesson 14**
**Lab Report**

The lab report is oral, consisting of pointing out and describing the chromosomes on the onion root tip slide and the *Ascaris* roundworm slide, your model of DNA, and the process of transcription.

• Award 5 possible points for each of these 3 steps

• Award an additional 5 possible points for carefully following directions and for making well-constructed DNA and RNA models

20 points possible overall.

**Lesson 15**
**Worksheet 1**

1. He found that in every case the amount of adenine was always the same as the amount of thymine and that the amount of guanine was always the same as the amount of cytosine.

2. 1

3. Codon

4. Amino, protein

5. B

6. A

7. C

**Lesson 15**
**Worksheet 2**

1. End

2. Beginning

3. End

4. Redundant

5. Grace

6. Alleles

7. Wild type

8. Recessive

9. Allele frequency

10. B

11. C

12. A

13. It is the same code that applies to the tentacles of an octopus and the tongue of an aardvark. If life evolved over millions of years, why has the code not varied?

**Lesson 15**
**Lab Report**

Explaining something to someone else is a powerful learning tool.

• Award 15 possible points for the student explaining to you the DNA and mRNA models, including how the mRNA is made from the DNA model

• Award an additional 5 possible points for answering the questions at the end of the assignment

Have the student explain the start and stop codons (groups of 3 bases that code for an amino acid in a resulting protein) in the model. Redundancies are different codons that result in the same amino acid in a protein. They do not have to include any redundancies in their model. Have the student explain to you what a redundancy is and describe one if it is included in the model. Have the student explain to you how changing a base could result in changing an amino acid that goes into a protein. This is an oral report rather than a written report. 20 points possible overall.

## Lesson 16
## Worksheet 1

1. Amino acids
2. Ribosomes
3. Transcription
4. Transcribe, nucleic acid (in this order)
5. D
6. A
7. B
8. C
9.

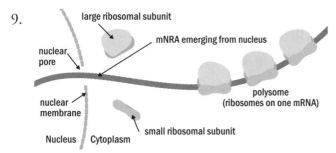

## Lesson 16
## Worksheet 2

1. Kind, species
2. Species
3. evolution
4. Enzymes
5. C
6. D
7. B
8. A
9. It is not that they have a greater knowledge of the data that keeps them away from God, it is their will and perspective on life. This means that

you do not have to check your brains out at the door to believe in God. It also means that their data are not compelling you to walk away from your faith. As you see in this book, there are many brilliant biologists whose faith is in Christ.

## Lesson 16
## Lab Report

Have your student demonstrate and explain to you how introns and exons work using a sentence (step 1) and a sequence of RNA bases (step 2). The student is to demonstrate to you the sequence of amino acids produced for 2 different patterns of introns and exons (step 2).

Have the student start with a sequence of RNA bases and demonstrate and explain a frameshift (step 3).

• Award a possible 7 points for each of steps 1 and 2
• Award a possible 6 points for step 3

20 points possible overall.

## Lesson 17
## Worksheet 1

1. The process of synthesizing proteins with the mRNA and ribosomes that is going from the language of nucleic acids to the language of proteins.
2. Ribosomes
3. Polysome
4. Transfer RNA, double
5. UCA
6. B
7. C
8. A

## Lesson 17
## Worksheet 2

1. B
2. D
3. A
4. C
5. Redundant, protein
6. mRNA, design

7. "writing out" a copy of a strand of the DNA as mRNA. It is in the same language as the codons of nucleic acids.

8. To write something in another language. Translation goes from the language of codons being the words in a nucleic acid to the language of amino acids being the words in proteins.

## Lesson 17
## Lab Report

The student is to orally point out from the prepared slides and describe to you the parts and functions of striated muscle cells, smooth muscle cells, and cardiac muscle cells as described in the assignment.

• Award a possible 5 points for each (15 points total)

The student is to orally point out from the prepared slide and describe to you the parts and functions of motor neurons as described in the assignment.

• Award a possible 5 points

20 points possible overall. The optional part of this lab is to be completed and graded next week.

## Lesson 18
## Worksheet 1

1. D
2. C
3. A
4. B
5. Homeobox, *Hox*
6. Body parts
7. Binary fission
8. Interphase
9. DNA, organelles
10. Mitosis
11. Cytokinesis
12. Prophase, nuclei, spindle fibers
13. The membranes around the nuclei come apart and microtubules called spindle fibers attach to the centromeres. At this time, the chromosomes become condensed, meaning that they become coiled and thicker and shorter; otherwise, they would be strung out very thin so that they could not be seen with the microscope.

## Lesson 18
## Worksheet 2

1. B
2. D
3. A
4. C
5. 4n, 2, 2
6. 46
7. Synapsis, tetrad
8. Loses, 1
9. Crossing over, survival
10. Cytoplasm, nourishment
11. Plasmids
12. Transformation
13. Transduction
14. Insulin
15. Genetic engineering
16. Recombinant
17. Before I formed you in the womb I knew you; Before you were born I sanctified you; I ordained you a prophet to the nations.

## Lesson 18
## Lab Report

The student is to point out the fern life cycle on the prepared slide and describe to you the stages and structures as given in the assignment. The structures are made up of different proteins made in the different stages. The student is to look up and describe to you some of the uses of ferns.

• Award 10 possible points for the description of the fern life cycle and uses

The student is to point out the moss life cycle on the prepared slide and describe to you the stages and structures as given in the assignment. The student is also to look up and describe to you some of the uses of mosses.

• Award 10 possible points for the description of the moss life cycle and uses

Have the student collect some ferns and mosses if they are available in your area. He/she is to get permission if they are in a neighbor's yard. If they are in a local nursery, you and the student can go and observe them without having to purchase them.

20 points possible overall.

## Optional Exercise

This exercise demonstrates the effects of different proteins (antibiotics which are products of mRNA translation) upon other organisms (bacteria).

The effects of antibiotics upon bacterial cultures are caused by their interactions with proteins on the surfaces or within bacterial cells. The student is to show and describe the results.

The student is to prepare a written report, with a chart showing the results of each of the 4 antibiotics upon the E. coli bacteria. The results are to be described in complete sentences.

- Award a possible 4 points for the effects of each of the four antibiotics, including how well the student followed directions and the neatness of the work and written report

- If the student did an extra sampling, add an additional possible 4 points to the grade, making the total possible 24 points

If this optional exercise is completed, it will be worth an additional 20 or 24 extra credit points. These 20 or 24 points will raise the total possible points at grading time by 20 or 24 points.

## Lesson 19
## Worksheet 1

1. Round

2. Wrinkled

3. Round

4. Round

5. Genotype

6. Phenotype

7.

|   | R | r |
|---|---|---|
| R | RR | Rr |
| r | Rr | rr |

**Punnett square of** $Rr \times Rr$

8. B

9. C

10. D

11. A

## Lesson 19
## Worksheet 2

1. C

2. A

3. B

4. Common, wild (in this order)

5. Dominant

6. $F^+F^f$, mildly frizzled

7. 25, 50, 25 (in this order)

8. Gregor Mendel noticed that plants produced from different crosses had properties in definite ratios indicating that they inherited definite objects.

9. He was able to predict the outcomes using the Punnett Square — a simple but very handy tool.

## Lesson 19
## Lab Report

The student is to prepare human pedigree charts for your family for each of tongue rolling, widow's peak hairline, freckles, blue or brown eyes, dimples, and PTC tasters. The directions for the charts are given in steps 1–4 of the assignment. The content of the charts depends upon your family characteristics.

- Award a possible 2 points for each chart. The charts with circles and squares demonstrate the physical traits (phenotypes) of each individual (parents, student, and siblings). This gives a total possible of 14 points.

The student is to prepare another chart for each of the pedigree charts. These charts are to show the genotypes. There are 2 copies of each gene with 1 being on each of 2 chromosomes. Each gene can be a dominant form or a recessive form. Each trait is produced by the combined effect of the 2 genes. The student is to determine from the physical traits the gene combinations and insert those into the new charts in place of the physical traits. As described in step 4, E is dominant and e is recessive for earlobes. Those with unattached earlobes can be EE or Ee. These are determined by the traits of the parents and offspring. For example, if one parent is Ee and the other is ee, the offspring cannot be EE. One gene is from one parent and the other is from the other parent. If this is confusing, have your student explain it. If you need to, feel free to ask a friend or relative to help. These are explained in chapter 19 of the text. For some of the pedigrees, there may not be enough information to determine the genotype. Those can be left blank.

- Award a possible 6 points for a good attempt at distinguishing the genotypes

20 points possible overall. This is a written report of the pedigree charts and genotype charts.

## Lesson 20
## Worksheet 1

1. Whichever allele is inherited for one gene does not affect which allele will be inherited for another gene (that is not linked).

2. C

3. A

4. B

5.

**Punnett Square** $RrYy$ x $RrYy$

|     | RY | Ry | rY | ry |
| --- | --- | --- | --- | --- |
| RY | $RRYY^1$ | $RRYy^1$ | $RrYY^1$ | $RrYy^1$ |
| Ry | $RRYy^1$ | $RRyy^2$ | $RrYy^1$ | $Rryy^2$ |
| rY | $RrYY^1$ | $RrYy^1$ | $rrYY^3$ | $rrYy^3$ |
| ry | $RrYy^1$ | $Rryy^2$ | $rrYy^3$ | $rryy^4$ |

6. Unpredictable

7. 5, significantly different (in this order)

## Lesson 20
## Worksheet 2

1. Linked, chromosome (in this order)

2. 3:1

3. 66%, 16%, 9%, 9% (in this order)

4. Crossing over

5. Synapse

6. Tetrad

7. Chromosomes

8. Karyotype

9. X, X

10. X, Y (in either order)

11. Genes, X, Y

12. Hemophilia, color blindness (in either order)

13. B

14. D

15. A

16. C

17. You were created in the image of God which gives you tremendous worth, no matter who you are. If you have accepted Christ's offer of forgiveness and restoration, then you can walk with Him in this life and in the future. Chemicals cannot do that.

## Lesson 20
## Lab Report

A written report is to be submitted with a chart showing the counts of the corn kernels as described in step 1.

- Award a possible 6 points for the chart. Is it neatly done, and does it make sense?

The student is to calculate the ratios of the kernel types by dividing the number of purple firm kernels by the number of yellow wrinkled kernels; the number of purple wrinkled kernels by the number of yellow wrinkled kernels and the number of yellow firm kernels by the number of yellow wrinkled kernels. For example, if there were 27 purple firm kernels and 3 yellow wrinkled kernels, 27/3 = 9. The expected total ratio is 9:3:3:1. The purple wrinkled divided by yellow wrinkled would be 3 and the yellow firm divided by the yellow wrinkled would also be 3. The student is to show the results and ratios from the calculations. It will not be exactly 9:3:3:1, but it should be close.

- Award a possible 6 points for completing the ratio. This part of the report is to be neatly organized and show the work that went into the calculations.

The report is to include answers to the questions in steps 4–7. The answers are to be in complete sentences — not just yes or no.

- The answer to each question is worth 2 possible points each (8 points total)

20 points possible overall.

**Lesson 21**
**Worksheet 1**

1.  generic and specific categories for classification
2.  Kinds
3.  Species, kind (in this order)
4.  John Ray, species (in this order)
5.  Carolus Linnaeus, species (in this order)
6.  Kind
7.  More
8.  B
9.  D
10. A
11. C

**Lesson 21**
**Worksheet 2**

1.  B
2.  A
3.  D
4.  C
5.  Plants, animals (in this order)
6.  Allopolyploid
7.  Flowers, fruit, seedless (in this order)
8.  Polydactyly
9.  Natural selection
10. Artificial selection
11. He was a Horticulture Professor at UCLA who produced the many varieties of roses that we have today by selectively breeding them (similar to the work of Gregor Mendel). He was also one of the founders of the Creation Research Society.

**Lesson 21**
**Lab Report**

This lab exercise is quite different from the others. The student is to organize a report around different orders of insects listed in the instructions. There are more orders of insects, but these present a good representation. The main goal of this exercise is to examine the variation in God's creation. There is variation of life forms within each kind of creation. Some of these variations are the result of mechanisms that were originally placed in DNA at creation. This is not evolution, because it is using mechanisms originally created acting upon DNA that was fully created in the first life forms.

The report should look like a well-organized listing of insects in each order. The student can consult insect books or go online to be sure that they are placing the insects in the proper order. Without doing breeding experiments with the insects (which is far beyond the scope of this exercise) it is hard to know for certain which would have descended from the same kind of creation. These are only suggestions on the part of the student.

In grading,

*   Assign a possible 10 points for thoroughness in including several insects in each order. They can include pictures if the resources to do so are readily available.
*   Assign a possible 10 points for neatness, logical organization, and the use of complete sentences in the descriptions

20 points possible overall.

**Lesson 22**
**Worksheet 1**

1.  that living organisms inherited actual objects that later became known as genes
2.  Bases, genes, chromosomes
3.  Computer software
4.  DNA Sequencing
5.  Treatments
6.  Base
7.  Single nucleotide polymorphisms
8.  Allele
9.  Genotype
10. B
11. D
12. C
13. A

## Lesson 22
### Worksheet 2

1. D
2. A
3. B
4. C
5. DNA, mature, nucleus
6. Sheep (lamb), Dolly, mature, egg, environment, problems
7. Aspen tree
8. Sea anemone
9. Answers will vary, but look for student's personal opinion regarding the cloning of humans in light of the author's view that cloning is dangerous and that cloned people would have a soul.

## Lesson 22
### Lab Report

There is no written report for this lab. The report is oral.

- Award a possible 5 points for pointing out to you the *Ranunculus* (buttercup flowering plant) root cross section epidermis, cortex, xylem tubes, and phloem tubes and their functions.
- Award a possible 5 points for pointing out to you the *Ranunculus* stem cross section epidermis, vascular tissue, and pith and their functions.
- Award a possible 5 points for pointing out to you the *Ficus* (ornamental tree) leaf cross section central vein, upper epidermis, mesophyll cells, lower epidermis, and the veins above the lower epidermis and their functions.
- Award a possible 5 points for completing step 5 as preparation for lab exercise 23.

20 points possible overall.

## Lesson 23
### Worksheet 1

1. D
2. A
3. B
4. C

5.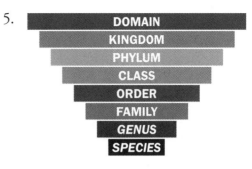

6. Kingdoms, Plantae
7. Phyla, classes, orders, families, genera, species
8. Capitalized, lowercased, underlined, italics
9. Genus, species
10. Vascular, non-vascular
11. Tubes
12. Seeds
13. Bryophytes
14. Gametophyte, gametes

## Lesson 23
### Worksheet 2

1. B
2. D
3. C
4. A
5. Mosses, 90
6. Sphenophyta
7. Hollow, silica
8. Pterophyta
9. Coniferophyta, Anthophyta (or Magnoliophyta)
10. Cones, needle-like
11. Evergreens, aspens
12. Monocotyledonae, Dicotyledonae
13. Grasses, lilies, corn, wheat, parallel, 3's
14. Fruits, beans, roses, sunflowers, branching, 4's
15. Drawings inspired by Genesis 1:11 will vary based on student's location.

## Lesson 23
## Lab Report

This report is part oral and part written.

- Award a possible 4 points for pointing out to you the outer plasma membrane, nucleus, cytoplasm, and pseudopod (if present) of the *Amoeba* and how it moves and digests food.

- Award a possible 4 points for pointing out to you the diverse diatoms and their role in their natural environment.

- Award a possible 4 points for pointing out to you a *Euglena* cell, its outer cell wall (because it is plant-like), nucleus, chloroplasts (green), and eyespot and their functions.

- Award a possible 4 points for pointing out to you a Volvox colony and its segments and its role as a photosynthetic algae.

- Award a possible 4 points for pointing out to you the mold sample that the student prepared, the yeast sample, and a written paragraph describing the structure and function of mold and yeast.

20 points possible overall.

## Lesson 24
## Worksheet 1

1. Claims were made that the salinity of water after the Flood would have kept plants from germinating after the Flood waters abated. Studies were done to see if many diverse plant seeds would germinate in seawater and they did.

2. Multicellular

3. Functions

4. Similarities, differences

5. John Ray, Carolus Linnaeus, order

6. Comparative anatomy

7. DNA, amino acid

8. Taxon

9. Taxon

10. Fossils

11. Cladistics

12. D

13. C

14. B

15. A

## Lesson 24
## Worksheet 2

1. The organism will die.

2. C

3. E

4. A

5. D

6. B

7. Mollusca, soft, mantle

8. Arthropoda, jointed appendage, exoskeleton

9. Arachnids, crustaceans, insects, centipedes, millipedes

10. Spiders, ticks, mites, horseshoe crabs, scorpions, 8, 2

11. Shrimp, lobsters, crabs, crayfish, barnacles, chelae, 2

12. Largest, 6, 3

13. Echinodermata, water vascular, sea water

14. Regenerate

15. Invertebrates, backbones, Chordata, notochord

## Lesson 24
## Lab Report

Grading this lab is as follows:

- Award a possible 4 points for your student pointing out and describing the *Planaria* and flatworms in general to you.

- Award a possible 8 points for your student showing you the dissected earthworm, and pointing out its structures and their functions to you. Have your student give you a mini-lecture on the value of earthworms as part of the 8 points.

- Award another possible 4 points based upon following directions and maintaining a clean workspace.

- Award a possible 4 points for pointing out to you the outer epidermis, muscle layers just under the epidermis, body cavity (coelom) outside of the intestine, intestine, typhlosole, and the cavity (lumen) inside the intestine of the earthworm cross section prepared slide.

20 points possible overall.

**Lesson 25**
**Worksheet 1**

1. Chordata, Vertebrata, Craniata
2. Notochord, dorsal tubular
3. Notochord
4. Pharynx, pharyngeal pouches
5. Gills, ear canals, larynx, parathyroid glands
6. Tail
7. Agnatha, Ostracoderms, armor plates
8. Hagfish, bony fish
9. Lampreys, bony fish
10. Chondrichthyes, sharks, skates, rays, scales, teeth
11. Oviparous, ovoviviparous, viviparous
12. Teeth
13. Osteichthyes, operculum
14. Gills, lungs
15.

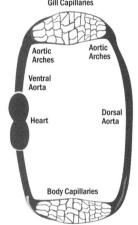

Fish circulation pattern

16.

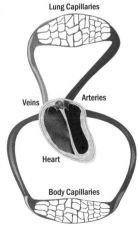

Land vertebrate circulation pattern

17. B
18. C

19. A

**Lesson 25**
**Worksheet 2**

1. Class Reptilia is a very diverse group. Some biologists feel that because reptiles differ so much from each other, the class reptilia should be divided into several classes. There are over 7,000 species of reptiles.
2. Salientia, leaping
3. Ectotherms, surroundings, poikilotherms, body temperature
4. Testudinata
5. Squamata
6. Crocodilia
7. Saurischia
8. Ornithischia
9. Aves, hollow
10. The fossil *Archaeopteryx* has been suggested as an indication of a reptile evolving into a bird. They were about the size of a crow. They had thecodont teeth (set in sockets in the jaw), no beak, cervical (neck) vertebrae that resemble a reptile more than a bird, small sternum (chest bone) that could not support flight muscles, and structures that resemble feathers. For many years it was considered to be a bird with reptile traits showing evolution from reptiles to birds, though now it is widely accepted to be simply a bird and not some evolutionary intermediate.
11. C
12. A
13. E
14. D
15. B

**Lesson 25**
**Lab Report**

Grading this lab is as follows:

• Award a possible 2 points for pointing out to you the cortex, medulla, developing eggs, and outer epidermis of the prepared slide of a frog ovary cross section.

- Award a possible 2 points for pointing out to you the cell body and flagellum of the prepared slide of frog sperm cells. The flagellum is a long whip-like tail that may be difficult to see because it is very narrow. You can award full credit even if the flagellum cannot be seen.

- Award a possible 12 points for your student showing you the dissected frog, and pointing out its structures and their functions. Have your student give you a mini-lecture on the value of frogs as part of the 12 points.

- Award another possible 4 points based upon following directions and maintaining a clean workspace.

20 points possible overall.

## Lesson 26
## Worksheet 1

1. In the beginning, God created living organisms to reproduce after their kind. Organisms reproduced and spread over the earth and many were later destroyed in the cataclysmic worldwide Flood. Fossils show evidence of mass extinctions where many organisms perished at once. Those that survived the Flood from the Ark of Noah spread over the earth until we have what we see today.

2. Kind

3. Small, existing DNA

4. Macroevolution

5. Holy Spirit, Creator

6. Earth

7. Flood Noah

8. Uniformitarianism

9. C

10. A

11. E

12. B

13. D

## Lesson 26
## Worksheet 2

1. *The Origin of the Species*, small changes, millions

2. Intermediate, not

3. 1900s, genetics, cytology, Agnostic Period

4. Punctuated equilibrium

5. Sedimentary, rapidly, slowly

6. Igneous, fossils

7. Metamorphic, pressures, temperatures, fossils

8. The geological column is a hypothetical picture of the fossil-bearing sedimentary rock layers. All of these layers do not appear anywhere on earth, but is the combination of many found in many different parts of the world. The uniformitarianist model assumes that these layers took long time periods to form. In contrast, a massive global Flood could cause most of the layers to form very rapidly. Heavier sediment would form the lower layers and lighter sediment would form the upper layers in the same way that sediment is deposited in streams, rivers, and lakes. This also explains why they do not all exist in one place. In each area, the sediment would form from what was available.

9. D

10. A

11. B

12. C

13. Features of the young earth creation view:

- The days of creation described in Genesis are considered literal 24-hour days

- The events of each day were as described in Genesis

- Adam and Eve were placed in the garden of Eden where they succumbed to the temptation from Satan to eat from the forbidden tree and were expelled from the garden

- Death entered through sin and the first death was of an animal used to cover the nakedness of Adam and Eve

- There was a Tower of Babel where God dispersed humans and changed their languages

- Noah constructed an Ark to save himself and his family from a worldwide Flood that destroyed all air breathing life that was not aboard the Ark and began reestablishing humans and animal life on the earth about six thousand years ago.

## Lesson 26
## Lab Report

Grading this lab is as follows:

- Award a possible 10 points for a written report on the student's findings of examples of bacterial decomposition as given in the lab instructions. This part of the report should be at least 3 to 4 pages long.

- Award a possible 5 points for a written report on the student's findings of examples of mold decomposition as given in the lab instructions. This part of the report should be at least 1 to 2 pages long.

- Award a possible 3 points for a written report on the student's findings of examples of lichen decomposition as given in the lab instructions. One page is sufficient for this part of the report. Lichen do not decompose dead organisms of waste but rather break down rock to produce sand in which other organisms can live.

- Award possible 2 points for neatness and clarity in the report.

In order for fossilization to occur, the specimen must somehow be prevented from decomposing. This is why without rapid burial, fossilization is a very rare event. This gives credibility to the rapid burial events of the Flood of Noah.

20 points possible overall.

### Optional Enrichment Exercise

Grading this enrichment exercise is as follows:

- Award a possible 10 points for the journal kept since lab exercise 10 describing the chicken leg fossil. Base the grade upon orderliness, thoroughness, and accuracy. Can you understand what is being described without the student's explanations?

- Award a possible 10 points for written answers to the following questions that are based upon the results and the journal.

Which is most likely to form a fossil — rapid burial or very slow burial where soil is deposited very little at a time?

What happens to a fossil in the ground?

How is a fossil different from the original bone structure?

Is all of an animal preserved as a fossil?

What does the fossil tell you about the original animal?

How would fossils have been buried in a flood?

Would a fossil form as well if it had wind gradually blowing sand over it without any wire covering?

If this optional exercise is completed, it will be worth an additional possible 20 extra credit points. These 20 points will raise the total possible points at grading time by 20 points.

## Lesson 27
## Worksheet 1

1. Occurred, patterns, true, repeated, seen
2. Slower, feathers, mammals
3. Convergent
4. Niches, variety, niche
5. Wing structure of bats

Wing structure of birds

6. B
7. A
8. E
9. D
10. C

## Lesson 27
## Worksheet 2

1. C
2. D
3. A
4. B

5. Vestigial, stand, crawl

6. 10%

7. Order, old, young, observable, measurable

8. Species, change, resembled, certain

9. Natural, design, species, kind

10. What was needed was to observe a smaller but powerful flood over a vast area and look at the results. The eruption of Mount St. Helens in 1980 helped in this regard. Even though it was on a much smaller scale than the Flood of Noah, it did demonstrate what a flood could do. It replaced opinion with observable measurable data. It caused many secular geologists to recognize that catastrophes are real and had to be considered. It also helped in understanding the limited variation possible within the kinds of creation.

## Lesson 27
## Lab Report

Grading this lab is as follows:

- Award a possible 10 points for the descriptions of various canids (dogs such as domestic dogs, wolves, coyotes, foxes, etc.) and which are likely to interbreed in their natural environment. There should be at least 10 different canids. The information on interbreeding is based more upon the student's background and readily available information. Do not reduce the grade because of limited information available. The purpose is to give suggestions and realize that there is much more research that could be done in this area. If they are known to be able to interbreed, they are in the same kind of creation. Consider neatness, grammar, and the use of good sentence structure.

- Award a possible 10 points for the description of various cats (domestic cats, tigers, lions, bobcats, etc.). There should be at least 10 different cats. Follow the same grading criteria as for the canids.

20 points possible overall.

## Lesson 28
## Worksheet 1

1. Yes and no. We have flesh and blood, but we are more than that. We have been created in God's image and given a soul as well as a body. We have flesh and blood bodies and are also created spiritually in the image of God Himself. Christ, who is the eternal God, took upon Himself human flesh so that He could die for us and redeem us to be with Him for eternity. To take on flesh, He allowed Himself to be born as we were, but without a human father. This means that His cells could have been haploid. When we have accepted Christ's redemptive work on our behalf, His Spirit is with us to carry us through this life. So, are we just another form of life like a gorilla? No. Even though we experience trials in this life so that we can grow to be more like Him, we have many physical qualities that we do not share with other life forms. We have intelligence, the ability to reason and speak, walk upright, and be creative with ideas and designs. This only begins to describe our unique qualities.

2. Domain: Eukarya

   Kingdom: Animalia

   Phylum: Chordata

   Class: Mammalia

   Order: Primates

   Family: Hominidae

   Genus: *Homo*

   Species: *sapiens*

3. Spiritual, image, God

4. Haploid, more vulnerable, mutant recessives

5. Share, intelligence, reason, speak, upright, ideas, designs

6. D

7. C

8. A

9. B

## Lesson 28
## Worksheet 2

1. Piltdown, hoax, human, lower jaw, ape

2. Cro-Magnon, *Homo sapiens*

3. Raymond Dart, *Australopithecus*, southern ape, brain, foramen magnum

4. *africanus*, *rhobustus*, *Homo habilis*, handyman, tools, *afarensis*, Lucy, humans, apes

5. Humans, apes, ancestors

6. C

7. E

8. B

9. A

10. D

11. This is shown in our origins and in our intimate relationship to our Creator who in His wisdom made us both physical and spiritual. We have a lot in common with other life forms because we were all created by Him. As you have seen, biological life is not simple. There are many enzyme systems directed by common DNA in all forms, from single cells to those with trillions of cells. If we had evolved, these should have changed over the years. Even in evolution theory this is acknowledged in what is called conserved traits. Again, remember, evolution is assumed, so they see no need to justify it. That also means that you do not have to feel compelled to adopt evolution because of it. Biological life is created by God and is good in His sight. He gave us through Adam the obligation to be stewards of His creation. By His grace, the similarities of other life forms to us has provided a blessing in being able to understand ourselves better by studying other animals.

## Lesson 28
## Lab Report

The student is to prepare a written report describing the following 5 areas based upon observations of the pictures made available in the instructions.

- Award a possible 4 points for each of the 5 areas. Grade on the basis of how well the available pictures were used, grammar, and sentence structure (20 points total).

Describe the similarities and differences between human skulls and chimpanzee (ape) skulls.

For each of the following, describe whether its skull looks human or apelike based upon your answer to question 1: Neanderthal, Cro-Magnon and *Australopithecus afarensis*.

Describe the similarities and differences between human skeletons and chimpanzee (ape) skeletons.

Does the Neanderthal skeleton look more human or ape-like?

What do you conclude about humans, chimpanzees, Neanderthals, Cro-Magnon, and *Australopithecus afarensis*?

Humans, Neanderthals, and Cro-Magnon have biped traits (walk on 2 legs). This includes being able to fully extend the knee and hip joints and have the opening to the skull (foramen magnum) at the bottom of the skull so that the head can rest on top of the body. Chimpanzees and the Australopithecines have quadruped traits (walk on 4 legs). These include not being able to fully extend the hip and knee joints so that if they tried to stand upright, they would be bent at the hip and knees. The opening (foramen magnum) in the skull where their spinal cord connects to the brain is toward the back of the skull so that the head would be positioned in front of the body facing forward.

# Biology Quiz Answer Keys

## Quiz #1
**Chemical Principles in Biology**

1. D
2. A
3. E
4. C
5. B
6. C
7. B
8. A
9. C
10. B
11. B
12. A
13. A
14. C
15. D

## Quiz #2
**Water**

1. D
2. B
3. B
4. D
5. C
6. A
7. C
8. B
9. D
10. A
11. B
12. C
13. D
14. A
15. E

## Quiz #3
**Carbohydrates and Lipids**

1. E
2. A
3. D
4. C
5. B
6. D
7. B
8. C
9. C
10. D
11. B
12. D
13. C
14. E
15. A

## Quiz #4
**Proteins and Nucleic Acids**

1. A
2. D
3. B
4. E
5. C
6. A
7. A
8. C
9. B
10. D
11. E
12. A
13. C
14. D
15. B

## Quiz #5
### Nature of Cells

1. D
2. A
3. E
4. B
5. C
6. D
7. C
8. A
9. B
10. E
11. C
12. B
13. D
14. E
15. A

## Quiz #6
### Cell Membranes and Nucleus

1. C
2. B
3. D
4. A
5. C
6. B
7. E
8. C
9. A
10. D
11. B
12. A
13. D
14. A
15. A

## Quiz #7
### Movement Through Cell Membranes

1. C
2. C
3. B
4. C
5. A
6. B
7. D
8. C
9. C
10. D
11. D
12. A
13. B
14. E
15. C

## Quiz #8
### Cell Organelles

1. E
2. C
3. A
4. D
5. B
6. C
7. B
8. A
9. D
10. C
11. C
12. A
13. D
14. E
15. B

**Quiz #9**
**Cell Division**

1. E
2. C
3. B
4. A
5. D
6. C
7. A
8. E
9. B
10. D
11. B
12. D
13. C
14. E
15. A

**Quiz #11**
**Biomes**

1. C
2. A
3. D
4. E
5. B
6. D
7. A
8. E
9. C
10. B
11. A
12. E
13. B
14. D
15. C

**Quiz #10**
**Ecosystems**

1. D
2. E
3. C
4. B
5. A
6. B
7. E
8. D
9. A
10. C
11. C
12. E
13. A
14. B
15. D

**Quiz #12**
**Energy Capture — Photosynthesis**

1. E
2. A
3. C
4. B
5. D
6. D
7. E
8. B
9. A
10. C
11. D
12. E
13. A
14. B
15. C

## Quiz #13
**Energy Release — Respiration**

1. C
2. D
3. B
4. A
5. E
6. B
7. E
8. A
9. D
10. C
11. B
12. D
13. A
14. E
15. C

## Quiz #14
**Chromosomes and Genes**

1. B
2. D
3. E
4. C
5. A
6. C
7. E
8. A
9. B
10. D
11. B
12. A
13. C
14. A
15. C

## Quiz #15
**The Genetic Code**

1. C
2. A
3. D
4. B
5. E
6. B
7. E
8. D
9. A
10. C
11. D
12. A
13. E
14. B
15. C

## Quiz #16
**Expression of DNA — Transcription**

1. D
2. B
3. E
4. A
5. C
6. E
7. D
8. A
9. C
10. B
11. C
12. D
13. A
14. E
15. B

## Quiz #17
### Expression of DNA — Translation

1. B
2. C
3. A
4. D
5. E
6. D
7. C
8. B
9. A
10. E
11. E
12. C
13. B
14. A
15. D

## Quiz #18
### Perpetuation of Life

1. B
2. C
3. D
4. E
5. A
6. C
7. D
8. A
9. B
10. E
11. B
12. E
13. A
14. C
15. D

## Quiz #19
### Genetic Patterns 1

1. D
2. C
3. A
4. B
5. E
6. B
7. D
8. E
9. C
10. A
11. A
12. C
13. D
14. B
15. E

## Quiz #20
### Genetic Patterns 2

1. D
2. E
3. C
4. B
5. A
6. C
7. A
8. E
9. B
10. D
11. D
12. C
13. B
14. E
15. A

## Quiz #21
### Genetic Mutations and Variations

1. B
2. D
3. A
4. E
5. C
6. E
7. B
8. A
9. C
10. D
11. C
12. D
13. E
14. A
15. B

## Quiz #22
### Genomics

1. B
2. D
3. E
4. A
5. C
6. E
7. A
8. D
9. C
10. B
11. E
12. B
13. A
14. C
15. D

## Quiz #23
### Plant Taxonomy

1. D
2. A
3. B
4. E
5. C
6. C
7. E
8. B
9. A
10. D
11. A
12. D
13. B
14. E
15. C

## Quiz #24
### Taxonomy — Invertebrates

1. D
2. A
3. E
4. B
5. C
6. A
7. C
8. B
9. E
10. D
11. E
12. C
13. D
14. A
15. B

**Quiz #25**
**Taxonomy — Vertebrates**

1. A
2. E
3. D
4. C
5. B
6. A
7. D
8. B
9. E
10. C
11. C
12. D
13. A
14. B
15. E

**Quiz #26**
**Views of Biological Origins**

1. D
2. E
3. A
4. C
5. B
6. B
7. E
8. C
9. A
10. D
11. A
12. D
13. E
14. C
15. B

**Quiz #27**
**Evidences of Biological Origins**

1. B
2. D
3. E
4. C
5. A
6. B
7. D
8. A
9. C
10. E
11. C
12. E
13. A
14. B
15. D

**Quiz #28**
**Human Origins**

1. B
2. E
3. D
4. C
5. A
6. B
7. D
8. E
9. A
10. C
11. B
12. D
13. A
14. C
15. E

# Biology —● Exam Answer Keys

**Examination #1 (Lessons 1–4)**

1. B
2. D
3. C
4. E
5. A
6. C
7. E
8. A
9. B
10. D
11. E
12. D
13. A
14. B
15. C
16. D
17. A
18. E
19. C
20. B
21. A
22. D
23. E
24. B
25. C
26. C
27. E
28. A
29. D
30. B

**Examination #2 (Lessons 5–8)**

1. E
2. C
3. A
4. D
5. B
6. D
7. C
8. E
9. B
10. A
11. B
12. A
13. D
14. E
15. C
16. C
17. D
18. A
19. B
20. E
21. C
22. B
23. E
24. D
25. A
26. A
27. D
28. B
29. E
30. C

## Examination #3 (Lessons 9–12)

1. B
2. D
3. E
4. C
5. A
6. D
7. A
8. E
9. C
10. B
11. C
12. E
13. A
14. B
15. D
16. B
17. A
18. D
19. E
20. C
21. E
22. C
23. A
24. D
25. B
26. B
27. C
28. E
29. A
30. D

## Examination #4 (Lessons 13–16)

1. B
2. C
3. A
4. E
5. D
6. E
7. D
8. C
9. A
10. B
11. D
12. C
13. E
14. B
15. A
16. C
17. D
18. B
19. E
20. A
21. C
22. E
23. B
24. A
25. D
26. D
27. C
28. B
29. A
30. E

**Examination #5 (Lessons 17–19)**

1. D
2. C
3. A
4. B
5. E
6. C
7. B
8. D
9. E
10. A
11. D
12. E
13. A
14. C
15. B
16. C
17. A
18. B
19. E
20. D
21. A
22. D
23. C
24. E
25. B
26. C
27. A
28. B
29. D
30. E

**Examination #6 (Lessons 20–22)**

1. B
2. D
3. E
4. C
5. A
6. B
7. A
8. E
9. C
10. D
11. C
12. E
13. D
14. B
15. A
16. B
17. A
18. C
19. E
20. D
21. D
22. C
23. A
24. B
25. E
26. D
27. A
28. C
29. B
30. E

**Examination #7 (Lessons 23–25)**

1. B
2. D
3. C
4. E
5. A
6. D
7. A
8. B
9. E
10. C
11. B
12. E
13. D
14. A
15. C
16. C
17. A
18. E
19. B
20. D
21. C
22. E
23. D
24. B
25. A
26. B
27. C
28. A
29. E
30. D

**Examination #8 (Lessons 26–28)**

1. C
2. B
3. E
4. A
5. D
6. C
7. A
8. E
9. D
10. B
11. D
12. B
13. C
14. E
15. A
16. B
17. D
18. A
19. C
20. E
21. C
22. A
23. E
24. D
25. B
26. D
27. A
28. B
29. E
30. C

# Biology Slide Cross-Reference

This reference provides page numbers in the *Master's Class Biology* curriculum and optional *Biology through a Microscope* book for the corresponding lab exercise slides. Additional information about the content in *Biology through a Microscope* is noted on the right.

## Lab 5

| Slide | Biology Student Book Pages | Biology through a Microscope Pages | Biology through a Microscope Notes |
|---|---|---|---|
| Thread | 54-55 | 6 | Darkfield microscopy contrast |
| Marker | Physical practice only, not imaged | | |

## Lab 6

| Slide | Biology Student Book Pages | Biology through a Microscope Pages | Biology through a Microscope Notes |
|---|---|---|---|
| Cheek Cells | 63 | 13 | Comparison of stains, not cheek cells |
| *Spirogyra* | 64 | 53 | |
| *Paramecium* | 64-65 | 84-85 | |
| Bacteria | 65 | 80-81 | Scanning electron microscope images |

## Lab 8

| Slide | Biology Student Book Pages | Biology through a Microscope Pages | Biology through a Microscope Notes |
|---|---|---|---|
| *Paramecium* | 81, 83 | 84-85 | |
| *Amoeba* | 84 | 78-79 | |
| Human Blood | 85 | | |
| Frog Blood | 85 | 19 | Darkfield microscopy contrast |

## Labs 9, 14

| Slide | Biology Student Book Pages | Biology through a Microscope Pages | Biology through a Microscope Notes |
|---|---|---|---|
| Onion Root Tip Mitosis | 89, 91 | | |
| *Ascaris* (Mitosis) | 93 | 28-29 | Large cross sections of whole organism |

## Lab 17

| Slide | Biology Student Book Pages | Biology through a Microscope Pages | Biology through a Microscope Notes |
|---|---|---|---|
| Muscle Types | 175-176 | 10 | Various magnifications |
| Motor Neurons | 176 | 10 | |

## Lab 18

| Slide | Biology Student Book Pages | Biology through a Microscope Pages | Biology through a Microscope Notes |
|---|---|---|---|
| Fern Life Cycle | 188 | 60-61 | |
| Moss Life Cycle | 188-189 | 70-71 | |

## Lab 22

| Slide | Biology Student Book Pages | Biology through a Microscope Pages | Biology through a Microscope Notes |
|---|---|---|---|
| *Ranunculus* root (c.s.) | 231 | 56-57 | |
| *Ranunculus* stem (c.s.) | 232 | 57 | |
| *Ficus* leaf (c.s) | 233 | Cover | |

## Lab 23

| Slide | Biology Student Book Pages | Biology through a Microscope Pages | Biology through a Microscope Notes |
|---|---|---|---|
| *Amoeba* | 546 | 78-79 | |
| Diatoms | 245-246 | 6, 53 | |
| *Euglena* | 245, 247 | 82-83 | Highlights pellicle |
| *Volvox* | 245-247 | 53 | |
| Yeast | | 62 | |
| Mold | | 62-63 | |

## Lab 24

| Slide | Biology Student Book Pages | Biology through a Microscope Pages | Biology through a Microscope Notes |
|---|---|---|---|
| *Planaria* | 260 | 16-17 | |
| Earthworm (c.s.) | 260 | | |

## Lab 25

| Slide | Biology Student Book Pages | Biology through a Microscope Pages | Biology through a Microscope Notes |
|---|---|---|---|
| Frog ovary | 279 | | |
| Frog sperm | 280 | | |
| Human skin | 280 | 11 | |